Walter Deane

Flora of the Blue Hills, Middlesex Fells, Stony Brook and Beaver Brook Reservations

Of the Metropolitan Park Commission, Massachusetts

Walter Deane

Flora of the Blue Hills, Middlesex Fells, Stony Brook and Beaver Brook Reservations
Of the Metropolitan Park Commission, Massachusetts

ISBN/EAN: 9783337272609

Printed in Europe, USA, Canada, Australia, Japan

Cover: Foto ©berggeist007 / pixelio.de

More available books at **www.hansebooks.com**

FLORA

OF THE

BLUE HILLS, MIDDLESEX FELLS, STONY BROOK AND BEAVER BROOK RESERVATIONS,

OF THE

METROPOLITAN PARK COMMISSION,

MASSACHUSETTS.

PRELIMINARY EDITION.

1896.

BOSTON:
C. M. BARROWS & CO.
1896.

To W. B. de las Casas, Esq.,

    Chairman Metropolitan Park Commission.

Sir: — In obedience to Orders of the Commission, dated January 26th and June 12th, 1895, and February 28th, March 11th, and May 6th, 1896, we hand you herewith the following preliminary list of the plants of the woodland reservations. This list is the product of volunteer work on the part of many botanists organized for this purpose by Mr. Warren H. Manning, lately a principal assistant in our office. The Commission, the public and the co-operating botanists are especially to be congratulated in that the list has been compiled and edited by Mr. Walter Deane.

    Respectfully submitted,

        OLMSTED, OLMSTED & ELIOT.

Brookline, Mass.,
May 15, 1896.

PREFACE.

The following list has been compiled at the request of the Metropolitan Park Commission in order to put on record the present condition of the vegetation of the new public reservations as a basis for comparison in the future. The acreage of Blue Hills Reservation is about 4,000 acres, Middlesex Fells Reservation 3,000 acres, Stony Brook Reservation 450 acres, Beaver Brook Reservation 58 acres. The diversified character of the lands embraced in the reservations gives rise to a most interesting flora. The height of Great Blue Hill is 635 feet, and throughout the reservations hills alternate with valleys and swamps, and clearings with woods, while ponds and brooks afford a rich aquatic vegetation. *Pogonia verticillata*, Nutt., *Habenaria fimbriata*, R. Br., *Epigæa repens*, L. and *Kalmia latifolia*, L. in the Blue Hills, and *Conopholis Americana*, Wallroth, in the Fells, are of the greatest interest, while in Monatiquot Stream which skirts the southern base of the Blue Hills are found the rare *Lemna Valdiviana*, Philippi, and the polymorphous *Sium Carsonii*, Durand. The public should be exhorted, if they come across such plants as these, to preserve them rigidly. The true botanist and lover of nature needs no such exhortation.

Direct botanical work in the reservations has been systematically prosecuted during the past two years only, and the record of plants is necessarily far from complete. This is especially true of Stony Brook and Beaver Brook Reservations. It has, nevertheless, been deemed best to publish the list at this time, in order that it may be used as a basis on which to build an exhaustive edition.

To enable me to compile the present list, the collections of the past two years, forming the nucleus of a Reservations Herbarium, have been placed in my hands, together with a large card catalogue of valuable notes by botanists who have studied the flora about Boston for years. Mr. M. L. Fernald kindly took charge of naming the Compositæ, and Dr. C. W. Swan the Junci, Cyperaceæ and Gramineæ. Dr. Swan has been much aided by Prof. F. Lamson-Scribner in the grasses, and Prof. L. H. Bailey in the genus Carex. I hold myself responsible for the rest of the Phanerogams, as I have examined them all with the exception of a few which were submitted to Dr. B. L. Robinson and Mr. M. L. Fernald. In the naming of these plants I have been assisted especially by the co-operation of Dr. G. G. Kennedy, Judge J. R. Churchill, Mr. L. L. Dame and Mr. E. F. Williams. Mr. George E. Davenport has taken charge of the Pteridophyta, Mr. E. L. Rand of the Bryophyta, Mr. J. W. Blankinship of the Characeæ, Mr. F. S. Collins of the Algæ, Miss Clara E. Cummings of the Lichens, and Prof. A. B. Seymour and Mrs. Flora W. Patterson of the Fungi. The list of Thallophytes is a small one, as but little work has as yet been done in this direction. Mr. E. L. Rand and Dr. B. L. Robinson have rendered most efficient assistance in regard to the form of the list, and Miss Mary A. Day, Librarian at the Gray Herbarium, has been most courteous in facilitating the work done in the Library. Mr. Warren H. Manning's large collections of plants, and his notes on the trees and shrubs have been of great service; and the general field notes of Mr. W. P. Rich, added to his extensive contributions to the herbarium, have been of especial value in regard to the habitat and distribution of the plants, particularly of the Fells. Judge J. R. Churchill, Dr. G. G. Kennedy, Messrs. C. E. Faxon, E. Faxon, L. L. Dame, F. S. Collins, E. F. Williams, H. A. Purdie, Dr. J. R. Webster, and Mrs. P. D. Richards have aided by their notes on the flora of the various reservations. To all the above and to those who have assisted by collecting in the reservations, contributing to the herbarium, or in any other way, my hearty thanks are extended.

Owing to the fact that this list is in reality a record of the plants growing in four separate reservations, a plan embracing clearness and brevity had to be considered. The list rests upon the authority of the Reservations Herbarium, various private herbaria, notably those of Messrs. C. E. and E. Faxon, W. P. Rich, F. S. Collins, E. F. Williams, Dr. G. G. Kennedy, Judge J. R. Churchill, the Gray Herbarium and my own, and lastly the reports of a few botanists, already mentioned, who have spent much time in studying the distribution of the plants of the reservations. I have also used my own field notes made during the past few years.

The names of plants indigenous to the reservations are printed in heavy full-faced type, while the names of plants introduced either into North America, or into the reservations from this country, or cultivated in old gardens, by roadsides and the like, are printed in small capitals. Synonyms are printed in italics.

After each species and its common name, if it has any, follows a statement of the general habitat of the plant as it occurs in New England. This ends with the period, separate observations being divided by semicolons. Then follow observations on each reservation. The following abbreviations for the reservations are used: *B*, for Blue Hills, *M*, for Middlesex Fells, *S*, for Stony Brook, *BB*, for Beaver Brook, and this order is followed, where possible. The other abbreviations needing a word of explanation are nat. for naturalized, adv. for adventive, int. for introduced, and cult. for cultivated. The letter indicating the reservation is followed in each case by the remarks on that reservation, separate remarks being divided by semicolons. The remarks on the different reservations are separated by a colon and dash. Every observation on the phanerogams made in the list is on record with the author's name and can be referred to in case of need. An asterisk following any letter standing for a reservation indicates that a specimen of the plant under discussion is found in the Reservations Herbarium; otherwise, the authority rests on private herbaria or on reliable reports.

The nomenclature is in general that of Gray's Manual, Sixth Edition, but in some cases a more recent revision of genera has been followed. In such cases the Manual synonymy, where possible, is given, and reference is made to the work containing the revision. With scarcely an exception the names of plants recorded as growing in old gardens are to be found in the last edition of Gray's Field, Forest and Garden Botany. It seems best to record these plants, as several old gardens are included in the reservations and it will be interesting to note the persistence or decay of the plants growing in them.

Let it be fully understood that this list is a preliminary one and that much work must be done in order to make it truly representative. The work of collecting and observing ought to be carried on with vigor so that a complete flora of the reservations may soon be forthcoming.

WALTER DEANE.

Cambridge, Mass.
May, 1896.

CATALOGUE OF PLANTS.

Series I. PHANEROGAMIA. FLOWERING PLANTS.

Class I. ANGIOSPERMÆ.

Sub-class I. DICOTYLEDONES.

Division I. POLYPETALÆ.

RANUNCULACEÆ. Crowfoot Family.

CLEMATIS, L. Virgin's Bower.

C. Virginiana, L.
 Moist ground, edges of thickets and roadsides, climbing over bushes. B^*, common : — M, frequent : — BB^*, occurs

ANEMONE, Tourn. Anemone. Wind-flower.

A. Virginiana, L.
 Open woods and meadows. B^*, Randolph Ave., etc. : — M^*, frequent on rocky wooded banks.

A. nemorosa, L.
 Open woods. B, M^* and BB, common : — S^*, occurs.

HEPATICA, Dill. Hepatica. Liver-leaf.

H. triloba, Chaix.
 Open woods and sunny banks. B, occasional; more common in the E. half of the Reservation : — M^*, frequent; Bear Hill; Cascade woods, etc.

ANEMONELLA, Spach.

A. thalictroides, Spach. RUE-ANEMONE.

Open woods and shady places. *B**, frequent: — *M**, frequent on rocky wooded hillsides : — *S**, dry woods, Bellevue Hill.

THALICTRUM, Tourn. MEADOW RUE.

T. dioicum, L. EARLY MEADOW RUE.

Rocky woods and shady roadsides. *B*, frequent : —*M**, common : — *S**, Overbrook Hill : — *BB**, occurs.

T. polygamum, Muhl. TALL MEADOW RUE.

Wet places, in sun and shade. *B**, common : — *M**, common in damp woods and meadows.

T. purpurascens, L. PURPLISH MEADOW RUE.

Dry uplands and rocky places. *B**, common, especially on the hill-tops : — *M**, rocky woods.

RANUNCULUS, Tourn. CROWFOOT. BUTTERCUP.

R. aquatilis, L. var. **trichophyllus**, GRAY. COMMON WHITE WATER-CROWFOOT.

Ponds and slow streams. *B**, occasional; Monatiquot Stream, etc. : — *M*, Spot Pond; stagnant pool.

R. multifidus, Pursh. YELLOW WATER-CROWFOOT.

Ponds and slow streams. *B*, rare; Monatiquot Stream : — *M*, frequent; stagnant pool, N. of Pine Hill; swamp, side of Bear Hill, etc.

R. abortivus, L. SMALL-FLOWERED CROWFOOT.

Shady places. *B* and *M**, common : — *BB**, by the brook.

Var. **micranthus**, Gray.

Shady places. *M**, Pine Hill, etc.

R. Allegheniensis, Britton.†

Similar in habitat and in aspect to R. *abortivus* and var. *micranthus*, but readily distinguished even in flower by the subulate hooked or recurved styles. *M**, damp rocky woods,

†See Bull. Torr. Bot. Club, xxii. 224.

N. of Wright's Pond, etc.: — *B B*, common by the brook in both sections.

R. recurvatus, Poir.   HOOKED CROWFOOT.
Open woods. *M**, damp shade, off Elm St.; valley, E. of Melrose Reservoir.

— **R. fascicularis,** Muhl.   EARLY CROWFOOT.
Dry slopes. *M**, abundant in open ground, summit of Bear Hill; W. border of N. Reservoir.

R. repens, L.   SPOTTED-LEAF BUTTERCUP.
Moist open ground. *B**, frequent: — *M* and *B B**, common.

R. BULBOSUS, L.   BULBOUS BUTTERCUP.
Nat. from Eu. in fields and by roadsides. *B*, *M* and *B B**, common.

— **R. ACRIS,** L.   TALL BUTTERCUP.
Nat. from Eu. in fields and by roadsides. *B*, *M* and *B B**, common: — *S*, frequent.

CALTHA, L.   MARSH MARIGOLD.

— **C. palustris,** L.
Swamps. *B*, *M** and *B B*, common: — *S*, frequent.

COPTIS, Salisb.   GOLDTHREAD.

C. trifolia, Salisb.   THREE-LEAVED GOLDTHREAD.
Damp woods. *B*, frequent; swamp, E. of Hawk Hill, etc.: — *M*, pine woods, W. of S. Reservoir.

AQUILEGIA, Tourn.   COLUMBINE.

— **A. Canadensis,** L.   WILD COLUMBINE.
Rocky shade. *B* and *M**, common: — *S**, Milkweed Hill: — *B B**, rare.

A. VULGARIS, L.   GARDEN COLUMBINE.
Cult. from Old World. *B*, persisting in old garden, Park's place, Hawk Hill.

DELPHINIUM, Tourn.   LARKSPUR.

D. CONSOLIDA, L.   FIELD LARKSPUR.
Cult. from Eu. and sparingly escaped. *M*, persisting for years in rubbish heap in woods, near Spot Pond.

ACTÆA, L. BANEBERRY.

A. spicata, L., var. rubra, Ait. RED BANEBERRY.
Rich woods. *M*, damp woods, near Spot Pond.

A. alba, Bigel. WHITE BANEBERRY.
Rich woods. *B**, frequent; Hawk Hill; Chickatawbut Hill, etc. :—*M**, Virginia Wood; E. of Melrose Reservoir.

PÆONIA, L. PEONY.

P. OFFICINALIS, Retz.
Cult. from Old World. *B*, old garden, Park's place, Hawk Hill.

BERBERIDACEÆ. BARBERRY FAMILY.

BERBERIS, L. BARBERRY.

B. VULGARIS, L. COMMON BARBERRY.
Nat. from Eu. in pastures and open places. *B*, *M* and *S*, common: — *B B**, frequent.

NYMPHÆACEÆ. WATER-LILY FAMILY.

BRASENIA, Schreb. WATER SHIELD.

B. peltata, Pursh.
Ponds and streams. *B*, occurs:—*M*, common in Spot Pond:—*S**, frequent in Turtle Pond.

NYMPHÆA, Tourn. WATER-LILY.

N. odorata, Ait. SWEET-SCENTED WATER-LILY.
Ponds and borders of streams. *B*, Monatiquot Stream:— *M*, common in Spot Pond:—*S*, common in Turtle Pond.

NUPHAR, Smith. COW-LILY. SPATTER DOCK. YELLOW POND-LILY.

N. advena, Ait. f.
Ponds and stagnant water. *B*, *M* and *S*, common:—*S**, a form near var. *variegatum*, Engelm., ditch, N. of Office.

SARRACENIACEÆ. Pitcher-plant Family.

SARRACENIA, Tourn. Pitcher-plant. Side-saddle Flower.

S. **purpurea**, L.
 Bogs and wet meadows. B^* and M, frequent: — S, occasional.

PAPAVERACEÆ. Poppy Family.

SANGUINARIA, Dill. Blood-root.

S. **Canadensis**, L.
 Rich woods. B^*, rare; Old Furnace Brook, etc.: — M^*, frequent; E. slope of Bear Hill; near Virginia Wood, etc.: — BB^*, common.

CHELIDONIUM, L. Celandine.

C. majus, L.
 Adv. from Eu. in waste places. M^*, frequent: — BB^*, common.

FUMARIACEÆ. Fumitory Family.

CORYDALIS, Vent.

C. **glauca**, Pursh. Pale Corydalis.
 Ledges and rocky places. B^*, M^* and S^*, common.

CRUCIFERÆ. Mustard Family.

CARDAMINE, Tourn. Bitter Cress.

C. **rhomboidea**, DC. Spring Cress.
 Wet meadows and damp woods. S^*. occasional.

C. **Pennsylvanica**, Muhl.† *C. hirsuta*, Gray, Man. ed. 6, 65, in part.
 Wet places. B^*, M and BB^*. common.

C. **parviflora**, L.† *C. hirsuta*, var. *sylvatica*, Gray, Man. ed. 6, 65.
 Rocky places. B, common: — M^* and S, frequent.

† See Syn. Fl. N. Amer. i. pt. 1, 158.

ARABIS, L. Rock Cress.

A. laevigata, Poir.
Rocky places. *M*, occasional; Cascade region; open woods, N. of Pine Hill, etc.

A. Canadensis, L. Sickle-pod.
Open woods. *B**, rare; Wild-cat Notch, etc.: — *M**, rare.

ALYSSUM, Tourn.

A. maritimum, L. Sweet Alyssum.
Cult. from Eu. *B*, spreading about beds and walks in old garden, Park's place, Hawk Hill.

NASTURTIUM, R. Br. Water Cress.

N. officinale, R. Br. True Water Cress.
Cult. from Eu. and escaped into brooks and ditches. *M*, occasional; ditch near Ravine Road, etc.: — *S**, Happy Valley: — *B B*, common.

N. palustre, DC. Marsh Cress.
Wet places. *M*, common.

BARBAREA, R. Br. Winter Cress.

B. vulgaris, R. Br., var. **arcuata**, Fries. Common Winter Cress. Yellow Rocket.
Wet places. *B*, frequent: — *M* and *B B**, occurs.

B. stricta, Andrz.† *B. vulgaris*, var. *stricta*, Gray, Man. ed. 6, 70.
Wet places. *M*, occurs: — *B B**, near Trapelo Road in N. section.

SISYMBRIUM, Tourn. Hedge Mustard.

S. officinale, Scop. Common Hedge Mustard.
Nat. from Eu. in waste places. *B* and *M*, common.

BRASSICA, Tourn.

B. Sinapistrum, Boiss. Charlock.
Adv. from Eu. in waste places. *S**, Washington St.

† See Syn. Fl. N. Amer. i. pt. 1, 150.

- B. NIGRA, Koch. BLACK MUSTARD.
Adv. from Eu. in waste places. *B*, occasional; Hawk Hill.
B. CAMPESTRIS, L. TURNIP.
Int. from Eu. and persisting in cult. fields and waste places. *M*, fields near Doleful Pond.
B. JUNCEA, Cosson.†
Int. from Eu. into waste places; generally glabrous; upper leaves nearly or quite entire, cuneate at the base. *B**, rare; old house-sites.

CAPSELLA, Medic. SHEPHERD'S PURSE.
- C. BURSA-PASTORIS, Moench.
Nat. from Eu. everywhere. *B* and *M*, common :—*S**, occasional :—*B B*, frequent.

LEPIDIUM, Tourn. PEPPERGRASS.
- L. Virginicum, L. WILD PEPPERGRASS.
Roadsides and waste places. *B** and *M*, common.

RAPHANUS, Tourn. RADISH.
- R. RAPHANISTRUM, L. WILD RADISH. JOINTED CHARLOCK.
Adv. from Eu. in fields and waste places. *B*, Randolph Ave.

RESEDACEÆ. MIGNONETTE FAMILY.

RESEDA, Tourn. MIGNONETTE.
R. ALBA, L. *R. suffruticulosa*, L. Sp. Pl. ed. 2, 645.
Adv. from Eu. in waste places. *B**, old garden, Park's place, Hawk Hill.

CISTACEÆ. ROCK-ROSE FAMILY.

HELIANTHEMUM, Tourn. ROCK-ROSE.
H. Canadense, Michx. FROST-WEED.
Dry soil in open ground. *B**, occasional.

† See Syn. Fl. N. Amer. i. pt. 1, 134.

LECHEA, Kalm. Pinweed.

L. major, Michx.
Dry sandy soil and sterile ground. B^*, frequent:—M, common in dry woods and fields:—S, occasional.

L. maritima, Leggett.†   *L. minor*, var. *maritima*, Gray, Man. ed. 6, 77.
Sandy soil. S, occurs.

L. intermedia, Leggett.‡   *L. minor*, Gray, Man. ed. 6, 77, in part.
Dry rocky ground. B^* and S, occurs:—M^*, frequent; border of Middle Reservoir, etc.

L. tenuifolia, Michx.
Dry sterile soil. B^*, common:—M^*, common by Middle Reservoir and in open woods:—S, occasional.

VIOLACEÆ. Violet Family.

VIOLA, Tourn. Violet.

V. pedata, L. Bird-foot Violet.
Sandy soil. B^* and S, frequent:—M^* and BB, common.

V. palmata, L., var. **cucullata,** Gray.
Low ground. B^* and M^*, common:—BB^*, occurs.

V. sagittata, Ait. Arrow-leaved Violet.
Open dry ground. B^* and M, common:—S^*, occurs.

V. blanda, Willd. Sweet White Violet.
Damp meadows and moist open woods. B^* and BB, frequent:—M^*, common:—S^*, swamp, S. W. of office.

V. primulæfolia, L. Primrose-leaved Violet.
Wet places. B^*, occasional; Pine Tree Brook meadow, etc.:—M^*, meadow, E. of Bear Hill:—BB^*, occasional.

V. lanceolata, L. Lance-leaved Violet.
Wet bogs and meadows. B^*, M^*, S^* and BB^*, common.

---

† See Syn. Fl. N. Amer. 1. pt. 1, 192.
‡ See Syn. Fl. N. Amer. 1. pt. 1, 193.

V. pubescens, Ait. Downy Yellow Violet.
Rich woods and shady places. *B**, frequent; near Hillside Pond; base of Wampatuck Hill, etc. :—*M*, Cascade woods; near Hemlock Pond.

Var. scabriuscula, Torr. & Gray.
Rich woods and shady places. *B B**, rare; by stone wall.

V. canina, L., var. Muhlenbergii, Gray. Dog Violet.
Damp shady places. *B*, occasional; Cragfoot Spring, etc. :—*M*, damp rocky woods, N. of Wright's Pond.

CARYOPHYLLACEÆ. Pink Family.

DIANTHUS, L. Pink.

D. Armeria, L. Deptford Pink.
Adv. from Eu. in waste ground. *S**, roadside, Washington St.

SAPONARIA, L.

S. officinalis, L. Bouncing Bet. Soapwort.
Adv. from Eu. in waste places and by roadsides. *B**, frequent.

SILENE, L. Catchfly. Campion.

S. Cucubalus, Wibel. Bladder Campion.
Nat. from Eu. in waste places. *B*, occurs.

S. Pennsylvanica, Michx. Wild Pink.
Sandy or gravelly places. *B B**, rare; W. of Waverly Oaks.

S. antirrhina, L. Sleepy Catchfly.
Waste places. *B*, common :— *M*, occasional; E. border of S. Reservoir; near top of Pine Hill, etc.

S. Armeria, L. Sweet William Catchfly.
Cult. from Eu.; weed, escaped. *B*, spreading in old garden, Park's place, Hawk Hill.

LYCHNIS, Tourn. Cockle.

L. Githago. Lam. Corn Cockle.
Adv. from Eu.; weed, especially in wheat-fields. *M**, wood-path, Stoneham.

L. Chalcedonica, L. Scarlet Lychnis.

Cult. from Old World and escaped. *B*, old garden, Park's place, Hawk Hill.

ARENARIA, L. Sandwort.

A. serpyllifolia, L. Thyme-leaved Sandwort.

Nat. from Eu. in sandy waste places. *B**, near Office: — *M**, occasional; Bear Hill; Winthrop Hill, etc.: — *BB**, about superintendent's house.

A. lateriflora, L.

Gravelly places, fields and thickets. *M*, occasional; low ground near Cascade, etc.: — *BB**, occurs.

STELLARIA, L. Chickweed. Starwort.

S. media, Smith. Common Chickweed.

Nat. from Eu. in gardens and damp places. *B* and *M*, common: — *S**, occasional.

S. graminea, L. English Starwort.

Adv. from Eu. in grassy places. *B*, frequent: — *M*, occasional: — *BB**, common, E. end of upper dam.

CERASTIUM, L. Mouse-ear Chickweed.

C. vulgatum, L.

Nat. from Eu. in fields, lawns, and waste places. *B* and *M*, common: — *S*, occasional: — *BB**, occurs.

BUDA, Adans. Sand-Spurrey.

B. rubra, Dumort.

Dry sandy or gravelly soil. *M*, common: — *S**, occurs.

PORTULACACEÆ. Purslane Family.

PORTULACA, Tourn. Purslane.

P. oleracea, L. Purslane. Pusley.

Nat. from Eu. in cult. and waste ground. *B** and *M*, common: — *S*, occurs.

ELATINACEÆ. Waterwort Family.

ELATINE, L. Waterwort.

E. Americana, Arn.
Margins of ponds and streams. M^*, abundant on margins of Spot Pond.

HYPERICACEÆ. St. John's-wort Family.

HYPERICUM, Tourn. St. John's-wort.

H. ellipticum, Hook.
Wet places. M^*, frequent.

H. perforatum, L. Common St. John's-wort.
Nat. from Eu. in fields and waste places. B^* and M, common: —S, frequent.

H. maculatum, Walt.
Damp shady places. B^*, occasional; near Hoosicwhisick Pond, etc.: — M^*, W. border of S. Reservoir.

H. mutilum, L.
Low ground. B^* and M^*, common.

H. Canadense, L.
Wet sandy soil. B^* and M^*, common.

H. nudicaule, Walt. Orange-grass. Pine-weed.
Sandy soil. B^* and M, common.

ELODES, Adans. Marsh St. John's-wort.

E. campanulata, Pursh.
Low swampy ground. B^* and M, frequent:—S^*, bog, W. of Turtle Pond.

MALVACEÆ. Mallow Family.

MALVA, L. Mallow.

M. rotundifolia, L. Common Mallow. Cheeses.
Nat. from Eu. in cult. ground and waste places. B^*, common:—M^* and S, occurs.

TILIACEÆ. Linden Family.

TILIA, Tourn. Linden. Basswood.

T Americana, L. Basswood.
Rich woods. B, comparatively rare; single, big, bushy tree on knoll in Pine Tree Brook valley; a few trees in Balster Brook valley and edge of upland, E. of Beech Run; Rattle Rock :—M, frequent; not uncommon as a small tree or shrub in rocky woods :—S, apparently rare; meadow, N. W. of Bold Knob, a small tree.

LINACEÆ. Flax Family.

LINUM, L. Flax.

L. Virginianum, L.
Open dry woods. B*, common :—M, frequent.

L. usitatissimum, L. Common Flax.
Adv. from Eu. in fields and waste places. B, road by Wolcott Pines.

GERANIACEÆ. Geranium Family.

GERANIUM, Tourn. Cranesbill.

G. maculatum, L. Wild Cranesbill.
Open woods, fields, and roadsides. B*, M and B B*, common :—S*, Bellevue Hill, etc., doubtless common.

G. Robertianum, L. Herb Robert.
Moist woods and shaded ravines. B*, frequent; Hancock Hill; Hancock valley, etc. :—M*, frequent on damp rocks in woods.

G. Carolinianum, L.
Waste places and barren soil. B, rare; Rattle Rock :—M*, occasional on rocks in open woods, and on upland rocks; Bear Hill, etc.

OXALIS, L. Wood Sorrel.

O. violacea, L. Violet Wood Sorrel.

Rich open woods, common South. *B B*, two plants in grass by the brook, 1883.

O. corniculata, L., var. **stricta,** Sav. Yellow Wood Sorrel.

Open ground and roadsides. *B**, common:—*M*, common in wood-paths and pastures.

IMPATIENS, L. Balsam Jewel Weed.

I. fulva, Nutt. Spotted Touch-me-not. Wild Balsam.

Open or shaded moist places. *B** and *M*, common:—*S*, bog in woods.

SIMARUBACEÆ. Quassia Family.

AILANTHUS, Desf. Chinese Sumach. Tree of Heaven.

A. glandulosus, Desf.

Cult. from China as a shade tree. *B*, old garden, Park's place, Hawk Hill.

RUTACEÆ. Rue Family.

XANTHOXYLUM, L. Prickly Ash.

X. Americanum, Mill.

Rocky woods and river banks. *B*, rare; near Hoosicwhisick Pond.

ILICINEÆ. Holly Family.

ILEX, L. Holly.

I. verticillata, Gray. Black Alder.

Low grounds and thickets; the bright-red berries are conspicuous in the autumn and are much used for Christmas decorations. *B**, *S* and *B B*, common on edges of bogs, pools, and open wet places, and sometimes in dry open ground;

easily killed back by fire, but sprouts from the old stumps and from the roots:—*M*, common; low grounds, borders of woods, and meadows.

I. lævigata, Gray. SMOOTH WINTERBERRY.
Wet ground. *S**, rare; border of Turtle Pond.

I. glabra, Gray. INKBERRY.
Sandy wet ground. *B**, occasional; Cedar Swamp; by Monatiquot Stream.

NEMOPANTHES, Raf. MOUNTAIN HOLLY.

N. fascicularis, Raf.
Swamps and damp woods. *B**, occasional; Cedar Swamp; Hemlock Pound, etc.: — *M*, occasional in wet places: — *S**, occasional; Happy Valley; by Turtle Pond, etc.

CELASTRACEÆ. STAFF-TREE FAMILY.

CELASTRUS, L. STAFF-TREE. SHRUBBY BITTER-SWEET.

C. scandens, L. WAX-WORK. CLIMBING BITTER-SWEET.
Dry rocky slopes and open woods. *B**, occasional; Hancock valley; Chickatawbut Hill, etc.: — *M*, frequent; slope of Bear Hill; woods by N. Reservoir, etc.: — *S**, woods, W. of Turtle Pond: — *B B**, occurs.

RHAMNACEÆ. BUCKTHORN FAMILY.

RHAMNUS, Tourn. BUCKTHORN.

R. cathartica, L. COMMON BUCKTHORN.
Cult. from Eu. for hedges, and nat. *B**, *M* and *S*, not generally distributed, but found in many places; pastures, hillsides, etc: — *B B**, frequent.

CEANOTHUS, L. NEW JERSEY TEA.

C. Americanus, L.
Dry open woods. *B**, common in shade on dry slopes and often in open fields: — *M**, common in open places in woods: — *S**, near Turtle Pond.

SIUM, L. Water Parsnip.

S. cicutæfolium, Gmelin.
In water and wet places. B^*, frequent: — M^*, common in meadows, ditches, and on borders of ponds.

S. Carsonii, Durand.
In running and still water; an interesting species, showing forms grading into the above. B^*, Monatiquot Stream.

CARUM, L. Caraway.

C. Carui, L.
Cult. from Eu. for the "caraway seed" and nat. in many places in fields and by roadsides. M, occasional.

CICUTA, L. Water Hemlock.

C. maculata, L.
Low grounds everywhere, very poisonous to taste. B^*, frequent: —M^*, frequent in meadows and by borders of ponds: —S, occurs.

C. bulbifera, L.
Wet places, very poisonous to taste. B^*, Cedar Swamp: — M, frequent; borders of N. and S. Reservoirs, etc.

OSMORRHIZA, Raf. Sweet Cicely.

O. brevistylis, DC.
Rich, moist, open woods. M, frequent in damp rocky woods; abundant at Cascade; Virginia Wood.

O. longistylis, DC.
Rich, moist, open woods. M, Virginia Wood; by Spot Pond: —BB^*, near The Falls.

HYDROCOTYLE, Tourn. Water Pennywort.

H. Americana, L.
Wet woods and meadows. B^*, occasional: — M, common: —S, occurs.

SANICULA, Tourn. BLACK SNAKEROOT.

S. Marylandica, L.
Thickets, open dry or moist woods, and meadows. *B**, common : — *M**, common in damp woods : — *S**, frequent in rocky woods.

ARALIACEÆ. GINSENG FAMILY.

ARALIA, Tourn. WILD SARSAPARILLA.

A. racemosa, L. SPIKENARD. LIFE-OF-MAN.
Rich soil in woods and open places. *B**, rare; W. slope of Great Blue Hill; Hawk Hill: — *M**, occasional.

A. hispida, Vent. BRISTLY SARSAPARILLA.
Rocky places and sandy soil. *B**, common; ledges, N. of Fox Hill, etc. :— *M*, common in upland rocky woods; abundant on S. shore of Spot Pond :—*S*, Milkweed Hill.

A. nudicaulis, L. WILD SARSAPARILLA.
Rich woods and moist woodlands. *B ** and *M*, common : — *B B**, occasional.

CORNACEÆ. DOGWOOD FAMILY.

CORNUS, Tourn. CORNEL. DOGWOOD.

C. Canadensis, L. BUNCHBERRY. DWARF CORNEL.
Damp woods. *B**, rare; valley, N. of Wampatuck Hill; Hawk Hill : — *M*, occasional : — *S**, occurs.

C. florida, L. FLOWERING DOGWOOD.
Dry and moist rocky woods. *B**, frequently found on dry and rocky slopes and near the edges of moist runs in the E. section, less frequently in the W. and central sections; easily killed by fire, but sprouts freely from the stumps : — *M**, occasional throughout; rather common in the region of Spot Pond : — *S**, frequent.

C. circinata, L'Her. ROUND-LEAVED CORNEL.
Rich rocky woods. *B**, ledges and woods; E. slope of

Hancock Hill; Great Dome, etc. : — M*, E. of Melrose Reservoir; rocky woods near Cascade: — S, occurs.

C. sericea, L. SILKY CORNEL. KINNIKINNIK.

Wet places. B* and M, common: — S, frequent: — BB*, occasional.

C. paniculata, L'Her. PANICLED CORNEL.

Dry open ground, roadsides, and thickets. B* and M*, common: — S*, rare; meadow, N. W. of Bold Knob; Overbrook Hill: — BB, occasional.

C. alternifolia, L. f. ALTERNATE-LEAVED CORNEL.

Woods, hillsides, banks of streams, and wet places. B and M, common: — S and BB, occasional in moist places.

NYSSA, L. TUPELO. SOUR-GUM TREE.

N. sylvatica, Marsh. TUPELO. BLACK OR SOUR GUM. PEPPERIDGE.

Swamps and rich woods. B, occasional; W. of Great Dome, etc.; the finest trees on W. side of Purgatory Road: — M, occasional; Great Island, etc. : — S*, occurs.

DIVISION II. GAMOPETALÆ.

CAPRIFOLIACEÆ. HONEYSUCKLE FAMILY.

SAMBUCUS, Tourn. ELDER.

S. Canadensis, L. COMMON ELDER.

Rich soil in open places; persists for a time in shady wet ground, but grows weaker each year and finally disappears. B and M, common: — S and BB, occasional.

S. racemosa, L. RED-BERRIED ELDER.

Rocky woods and shaded slopes. B, rare.

VIBURNUM, L. ARROW WOOD.

V. acerifolium, L. MAPLE-LEAVED VIBURNUM. DOCKMACKIE.

Rocky woods; much injured by fires, but spreads freely from the roots. B^3, M and S*, common.

V. dentatum, L. Arrow Wood.

Swamps, meadows, damp woods, and roadsides. *B** and *M*, common:—*S** and *B B**, occasional.

V. cassinoides, L. Withe-rod.

Moist woods and wet places. *B**, frequent; Hawk Hill; Purgatory Road; Blueberry Swamp; Bouncing Brook valley, etc. :—*M* and *S**, occasional in swampy places.

V. Lentago, L. Sweet Viburnum.

Open woods, banks of streams, and clearings. *B**, occasional; W. of Pasture Run; near Streamside Ledge; Chickatawbut Hill, etc. :—*M**, frequent:—*S**, Happy Valley:—*B B**, occasional.

TRIOSTEUM, L. Feverwort. Horse Gentian.

T. perfoliatum, L.

Rich woods. *B*, occasional; S. side of Great Blue Hill; Purgatory Road, etc. :—*M*, occasional.

LONICERA, L. Honeysuckle.

L. sempervirens, Ait. Trumpet Honeysuckle.

Wild from N. Y. south, common in cultivation. *B*, escaped:—*M*, escaped into the woods, very local.

L. Japonica, Thunb. Japanese Honeysuckle.

Cult. from Japan and China. *B*, old garden, Park's place, Hawk Hill.

DIERVILLA, Tourn. Bush Honeysuckle.

D. trifida, Moench.

Rocky open places. *B **, common on cliffs, ledges, and hill-tops :—*M*, common in rocky open woods :—*S* and *B B*, frequent.

RUBIACEÆ. Madder Family.

HOUSTONIA, L.

H. cærulea, L. Innocence. Bluets. Quaker Ladies.

Open, moist, grassy places. *B* and *M*, common: — *S*, frequent.

H. purpurea, L., var. **longifolia,** Gray.
Rocky or gravelly soil in open places. *B**, common: — *M**, occasional on rocks in upland woods; Bear Hill; E. border of N. Reservoir, etc.: — *S **, slope of Bellevue Hill, etc.

CEPHALANTHUS, L. Button Bush.

C. occidentalis, L.
Swamps and wet places, not persisting long in heavy shade or dry soil. *B **, common: — *M**, *S* and *B B*, frequent.

MITCHELLA, L. Partridge Berry.

M. repens, L.
Rich woods, creeping. *B **, *M **, *S ** and *B B*, common.

GALIUM, L. Bedstraw.

G. verum, L. Yellow Bedstraw.
Nat. from Eu. in dry places. *S **, occasional; covering a knoll, Happy Valley; Washington St.:—*B B **, a single clump.

G. Aparine, L. Cleavers.
Moist shady places. *M*, frequent in low grounds and damp woods.

G. pilosum, Ait.
Dry rocky woods. *B **, occasional: — *M **, rare.

G. circæzans, Michx. Wild Liquorice.
Dry rocky woods. *B* and *M **, frequent.

G. lanceolatum, Torr. Wild Liquorice.
Dry rocky woods. *B **, common: — *M **, frequent: — *S **, Overbrook Hill.

G. boreale, L. Northern Bedstraw.
Rocky banks of streams, westward; uncommon in New England, but occasionally planted. *B*, old garden, Park's place, Hawk Hill.

G. trifidum, L. SMALL BEDSTRAW.

Bogs, meadows, and moist open woods. $B*$ and $M*$, common : — S and BB, occurs : — $M*$ and $S*$, forms approaching var. *latifolium,* Torr.

Var. latifolium, Torr.

Same habitat as the species. $M*$, rare.

G. asprellum, Michx. ROUGH BEDSTRAW.

Low thickets, roadsides, and wet places. $B*$, Cedar Swamp : — M, frequent in low ground on borders of meadows and ponds.

G. triflorum, Michx. SWEET-SCENTED BEDSTRAW.

Open and rocky woods, and boggy places. $B*$, occasional; Hawk Hill, etc. : — $M*$, occasional; near Cascade and S. Reservoir, etc.

COMPOSITÆ. COMPOSITE FAMILY.

MIKANIA, Willd. CLIMBING HEMP WEED.

M. scandens, L.

Low thickets by streams. $B*$ and $S*$, occasional : — $BB*$, abundant by the brook just N. of Trapelo Road.

EUPATORIUM, Tourn. THOROUGHWORT.

E. purpureum, L. JOE-PYE WEED.

Meadows, swamps, and wet places. $B*$, $M*$, S and BB, common.

Var. maculatum, Darl.†

Same habitat as the species. $B*$, road to Sawcut Notch.

Var. amœnum, Gray.

Dry woods. $B*$, rare.

E. teucrifolium, Willd.

Low woods and moist thickets. $B*$, frequent : — $M*$, woods, W. border of Spot Pond : — $S*$, border of thickets by Turtle Pond.

† See Syn. Fl. N. Amer. i. pt. 2, 96.

E. rotundifolium, L., var. **ovatum,** Torr.

Low grounds. *B**, occasional; valley, N. of Wampatuck Hill; slope of Buck Hill, etc.

E. sessilifolium, L. UPLAND BONESET.

Dry wooded ground. *B**, occasional; ledges on slopes of Great Blue Hill; Babel Rock, etc.

E. perfoliatum, L.

Low shady ground. *B** and *M*, common: — *S*, damp woods.

E. ageratoides, L. WHITE SNAKE-ROOT.

Rich woods. *M**, rare.

E. aromaticum, L.

Dry woods. *B**, occasional; S. slope of Great Blue Hill; W. slope of Hancock Hill; Hawk Hill: — *M*, W. shore of Spot Pond, etc.: — *S**, rocky hillside by Turtle Pond.

SOLIDAGO, L. GOLDEN ROD.

S. cæsia, L.

Open rich woods and shaded roadsides. *B*, frequent: — *M* and *S*, common in damp woods.

S. latifolia, L.

Moist woods and shaded places. *B**, occasional; ledges, N. slopes of Wampatuck and Fox Hills: — *M**, occasional; Cascade, etc.

S. bicolor, L.

Dry woods, fields and roadsides. *B**, *M* and *S*, common.

S. puberula, Nutt.

Sandy ground. *B*, occasional: — *M*, open thickets.

S. odora, Ait. SWEET GOLDEN ROD.

Dry or sandy soil. *B**, common: — *M**, frequent on borders of woods and by dry woody paths: — *S**, dry woods.

S. rugosa, Mill.

Moist or dry ground, fields, and edges of meadows. *B** and *M*, common: — *S*, frequent.

S. ulmifolia, Muhl.

Moist woodlands and fields. $B*$, occasional; by Hoosic-whisick Pond; by Cedar Swamp, etc. : — $M*$, occasional; damp woods, W. borders of Spot Pond, etc. : — $S*$, rocky woods, etc.

S. Elliottii, Torr. & Gray.

Swamps and meadows. $B*$, rare.

S. neglecta, Torr. & Gray.

Swamps, especially sphagnous bogs. B, occasional; by Monatiquot Stream, etc. : — M, borders of N. Reservoir: — S, a form approaching var. *linoides*, Gray, border of Turtle Pond.

Var. linoides, Gray.

Same habitat as the species. S, rare.

S. arguta, Ait.

Moist woods. $B*$, rare : — M, frequent on Pine Hill.

S. juncea, Ait. EARLY GOLDEN ROD.

Dry rocky ground and open woods. $B*$ and $M*$, common.

S. serotina, Ait.

Moist or rich ground. M, frequent.

Var. gigantea, Gray.

Moist or rich ground. B, common : — $M*$, meadow, S. E. of Bear Hill : — $S*$, swamp.

S. Canadensis, L.

Meadows, fields, and shady ground. B, M, S and BB, common.

Var. glabrata, Porter.†

Same habitat as the species. $B*$, rare.

S. nemoralis, Ait.

Dry sterile fields and open woods. $B*$ and $M*$, common :— S, frequent.

S. lanceolata, L.

Moist ground, fields, roadsides, and borders of woods. $B*$, frequent : — M, common.

† See Bull. Torr. Bot. Club, xxi. 310.

SERICOCARPUS, Nees. WHITE-TOPPED ASTER.

S. conyzoides, Nees.
Dry open ground. *B**, common on Hawk Hill, Hancock Hill, Great Blue Hill, etc. : — *M**, common in rocky woods and abundant on rocky borders of Reservoirs : — *S**, occurs.

S. solidagineus, Nees.
Dry or moist woods and hill-tops. *B**, frequent on or near the exposed summits of the hills ; more common, apparently, W. of Hillside St. ; Great Blue Hill ; Hancock Hill ; Boyce Hill, etc. : — *M**, rare ; dry woods.

ASTER, L. ASTER.

A. corymbosus, Ait.
Dry woods. *B**, common : — *M*, frequent.

A. macrophyllus, L. GREAT-LEAVED ASTER.
Moist woods. *B**, common, not blossoming every year : — *M**, occasional in rocky woods in Cascade region, etc. : — *S**, low ground near Turtle Pond.

A. Herveyi, Gray.
Borders of oak woods in company with *A. macrophyllus* and *A. spectabilis*, and possessing characteristics of both species. *B*, rare ; Great Blue Hill : — *S*, rare.

A. spectabilis, Ait.
Sandy soil. *B**, rare ; slope of Great Blue Hill ; near Rattlesnake Hill, etc. : — *S*, rare.

A. Novæ-Angliæ, L.
Low ground. *B**, rare ; base of Great Blue Hill, etc.

A. patens, Ait.
Dry open grounds. *B**, *M* and *S**, common in dry woods.

A. undulatus, L.
Dry ground, borders of woods, and roadsides. *B*, *M* and *S*, common.

A. cordifolius, L. HEART-LEAVED ASTER.
Wooded banks and open woods. *B*, frequent : —*M*, common in rocky woods.

A. lævis, L.

Borders of woods, roadsides, and fields, in rather dry ground. *B*, frequent:—*M*, common.

A. ericoides, L.

Dry open places. *M*, rare; near N. Reservoir.

A. multiflorus, Ait.

Dry open ground or sterile soil. *B*, common:—*M*, N. border of N. Reservoir.

A. dumosus, L.

Borders of moist or dry woods. *B**, occasional; W. of Babel Rock, etc.:—*M**, W. border of S. Reservoir.

A. vimineus, Lam.

Moist ground. *B*, common:—*M*, common in meadows and low grounds.

Var. foliolosus, Gray.

Same habitat as the species. *B**, base of Great Dome; road up E. slope of Shadow Point.

A. diffusus, Ait.

Dry or moist open ground. *B*, common:—*M* and *S*, frequent.

A. paniculatus, Lam.

Moist, generally shady ground. *M**, meadows and ditches round Spot Pond; meadow, S. of Bear Hill.

A. longifolius, Lam.

Low ground. *B*, rare; N. E. part of Reservation.

A. Novi-Belgii, L.

Low or wet ground. *B* and *M*, common.

A. tardiflorus, L.

Low ground. *B*, rare; N. E. part of Reservation.

A. puniceus, L.

Swamps, low thickets, damp roadsides, and meadows. *B* and *M*, frequent.

Var. lævicaulis, Gray.

Same habitat as the species. *B*, rare; Cedar Swamp.

A. umbellatus, Mill.

Meadows, moist thickets, and roadsides. B^* and S, common: —M^*, frequent.

A. linariifolius, L.

Dry sandy or gravelly soil, in open places. B^*, M^* and S, common.

A. acuminatus, Michx.

Damp woods and shaded meadows. B^*, frequent: —M^*, common: —S, occurs.

A. nemoralis, Ait.

Bogs and swamps. B^*, rare; shore of Hoosicwhisick Pond; not typical, but approaching var. *Blakei,* Porter.†

ERIGERON, L. FLEABANE.

E. Canadensis, L. HORSEWEED. BUTTERWEED.

Waste places, fields, and roadsides; a common weed. B^* and BB, frequent: —M, common: —S, occasional.

E. annuus, Pers. DAISY FLEABANE.

Fields and waste places. B, occurs: —M, frequent.

E. strigosus, Muhl. SMALLER DAISY FLEABANE.

Fields and waste places. B^* and M, common: —S^*, Bearberry Hill, etc.: —BB^*, occurs.

E. bellidifolius, Muhl. ROBIN'S PLANTAIN.

Damp ground in fields and open woods. B^*, occasional: — M, frequent: —BB, occurs.

E. Philadelphicus, L. COMMON FLEABANE.

Moist open woods and fields. B, slope of Great Blue Hill: —S^*, occurs.

ANTENNARIA, Gaertn. EVERLASTING.

A. plantaginifolia, Hook. PLANTAIN-LEAVED EVERLASTING.

Dry open or shaded ground. B, common: —M^*, common in open rocky woods, etc.: —S^* and BB, frequent in thin poor soil.

† See Bull. Torr. Bot. Club, xxi. 311.

ANAPHALIS, DC. Everlasting.

A. margaritacea, Benth. & Hook. Pearly Everlasting.

Open woods, dry fields, and roadsides. B^*, common : — M^*, frequent in dry woods and on rocky borders of Reservoirs :— S^*, frequent ; S. slope of Bellevue Hill.

GNAPHALIUM, L. Cudweed.

G. polycephalum, Michx. Common Everlasting.

Dry open woods, fields, and roadsides. B^*, occasional : — M^*, frequent on rocky borders of Winchester Reservoirs and Spot Pond, etc. : —S, dry rocks.

G. decurrens, Ives. Everlasting.

Dry open ground. B^*, rare ; ledge on slope of Fox Hill.

G. uliginosum, L. Low Cudweed.

Low ground. B^*, M^* and S, common.

AMBROSIA, Tourn. Ragweed.

A. artemisiæfolia, L. Roman Wormwood. Ragweed.

Dry cult. and waste ground. B^* and M, common : — S, frequent : — BB, occasional.

XANTHIUM, Tourn. Cocklebur.

X. Canadense, Mill., var. **echinatum**, Gray.

Sandy soil near the sea. B, house-site, base of Rattlesnake Hill : — M, along Highland Ave. and Elm St.

RUDBECKIA, L. Cone Flower.

R. hirta, L. Black-eyed Susan.

Nat. from the West in meadows and fields. B^*, near old house-sites : — M, frequent in cult. ground.

HELIANTHUS, L. Sunflower.

H. annuus, L. Common Sunflower.

Cult. from the West and run wild in waste places. M, old house-site.

H. divaricatus, L.

Dry open ground and edges of thickets. *B**, common: — *M*, dry woods, base of Pine Hill, etc.: — *S**, dry woods near Turtle Pond, etc.

H. strumosus, L.

Open woods. *B*, rare.

BIDENS, L. Bur Marigold.

B. frondosa, L. Common Beggar-ticks.

Wet open ground, meadows, and damp roadsides. *B** and *M**, common: — *S** and *BB*, occasional.

B. connata, Muhl. Swamp Beggar-ticks.

Wet open ground. *B**, frequent; Pine Tree Brook meadow, etc.: — *M**, common in ditches and meadows round Spot Pond: — *S**, bog.

B. cernua, L. Smaller Bur Marigold.

Wet places. *M**, abundant in ditches and meadows round Spot Pond, etc.: — *BB**, occurs.

B. Beckii, Torr. Water Marigold.

Ponds and slow streams. *B**, rare; Monatiquot Stream.

ANTHEMIS, L. Chamomile.

A. Cotula, DC. Mayweed.

Nat. from Eu. by roadsides and in all waste places. *B*, common: —*M*, *S* and *BB*, occasional.

ACHILLEA, L. Yarrow.

A. Millefolium, L. Common Yarrow.

Dry grassy fields and roadsides. *B*, common; Hawk Hill; Hancock Hill, etc.: —*M*, common in wood-roads, paths, fields, and open rocky woods; the pink form, near N. Reservoir: —*S** and *BB*, common.

A. Ptarmica, L. Sneezewort.

Cult. from Eu. *B*, old garden, Park's place, Hawk Hill.

CHRYSANTHEMUM, Tourn.   Ox-eye Daisy.

C. Leucanthemum, L.   Daisy. White-weed.

Nat. from Eu. as a common weed in pastures, meadows, and by roadsides. *B, M* and *BB*, common : —*S*, occasional.

C. Parthenium, Pers.   Feverfew.

Cult. from Eu. and adv. in some places. *B**, spreading as a weed in old garden, Park's place, Hawk Hill.

TANACETUM, L.   Tansy.

T. vulgare, L.   Common Tansy.

Nat. from Eu. by roadsides, old dwellings, and escaping into fields. *B**, common : —*M* and *S**, occasional.

ARTEMISIA, L.   Wormwood.

A. vulgaris, L.   Mugwort.

Adv. from Eu. in waste places near dwellings. *B*, occasional; Willard St.

A. biennis, Willd.

Dry open places; native West, and becoming nat. in the East. *B**, rare; old cult. field near Hemlock Bound.

SENECIO, Tourn.   Groundsel.

S. aureus, L.   Golden Ragwort.

Swamps, meadows, and bogs. *B*, frequent : —*M*, common : —*S**, occasional.

Var. Balsamitæ, Torr. & Gray.

Rocky open woods. *B**, occasional; near summit of Great Blue Hill : —*M**, rare; Bear Hill.

ERECHTITES, Raf.   Fireweed.

E. hieracifolia, Raf.

Woods and thickets, becoming very vigorous on recently burnt ground. *B** and *M*, common : —*S*, frequent.

Forma, *E. ambigua*, DC.†, a smaller plant with petioled leaves; same habitat as the species. *B*, rare : —*M**, rare; near S. Reservoir.

† See DC. Prodr. vi. 295.

ARCTIUM, L. BURDOCK.

A. LAPPA, L., var. MINUS, Gray.
Nat. from Eu. in all waste places; our common form. *B, M, S* and *B B*, occasional.

CNICUS, Tourn. THISTLE.

C. LANCEOLATUS, Hoffm. COMMON THISTLE.
Nat. from Eu. in pastures, waste places, and by roadsides. *B**, meadow by Pine Tree Brook:—*M*, common, especially along roadsides and in cult. ground; border of swamp, E. side of Bear Hill.

C. muticus, Pursh. SWAMP THISTLE.
Swamps and low ground. *B*, occasional.

C. pumilus, Torr. PASTURE THISTLE.
Dry pastures and roadsides. *B**, occasional; Pine Tree Brook valley, etc.:—*M**, frequent in open places; woods on slope of Bear Hill, etc.

C. ARVENSIS, Hoffm. CANADA THISTLE.
Nat. from Eu. as a common and often troublesome weed in pastures, cult. fields, and by roadsides. *M*, frequent: — *S*, occasional.

CENTAUREA, L. STAR THISTLE.

C. CYANUS, L. BACHELOR'S BUTTON. CORNFLOWER.
Cult. from Eu. and escaped from gardens to roadsides and waste places. *B*, persisting from seed in old garden, Park's place, Hawk Hill.

C. NIGRA, L. KNAPWEED.
Adv. from Eu. in fields, waste places, and by roadsides. *M**, well established on Forest St. and near N. Reservoir; S. shore of Spot Pond, etc.

KRIGIA, Schreb. DWARF DANDELION.

K. Virginica, Willd.
Dry pastures and sandy places. *B**, common, especially on

hill-tops: — *M*, common on upland exposed rocks: — *S* *, occasional; Milkweed Hill, etc.

CICHORIUM, Tourn. CHICORY. SUCCORY.

C. INTYBUS, L.

Nat. from Eu. in fields and by roadsides. *B* *, field by Old Houghton Place: — *M*, common: — *S*, about house-site, Happy Valley.

LEONTODON, L. HAWKBIT.

L. AUTUMNALIS, L. FALL DANDELION.

Nat. from Eu. in fields, meadows, and by roadsides. *B* * and *M*, common.

HIERACIUM, Tourn. HAWKWEED.

H. Canadense, Michx.

Dry open woods and roadsides. *B* *, rare; observed in two localities.

H. paniculatum, L.

Dry open woods and shaded roadsides. *B* * and *M* *, common: — *S*, occasional.

H. venosum, L. RATTLESNAKE-WEED.

Dry open woods. *B* *, *M* * and *S* *, common.

H. Marianum, Willd.

Dry open woods. *B*, rather common: — *M*, rare; near Bear's Den Path: — *S* *, rare; Milkweed Hill.

H. scabrum, Michx.

Dry open woods and shaded roadsides. *B* *, *M* * and *S* *, frequent.

PRENANTHES, Vaill. RATTLESNAKE-ROOT.

P. alba, L. WHITE LETTUCE. RATTLESNAKE-ROOT.

Rich open woods. *M* *, common in damp woods: — *S*, woods near Turtle Pond.

P..serpentaria, Pursh.
Dry soil in open places. *B**, common: — *M** and *S**, occasional.

TARAXACUM, Haller. DANDELION.

T. OFFICINALE, Web. COMMON DANDELION.
Int. from Eu. into pastures, fields, and by roadsides everywhere. *B*, *M**, *S* and *B B*, common.

T. ERYTHROSPERMUM, Andrz.†
Native to Unalaska, etc., and int. into E. New England; readily distinguished from *T. officinale* by its smaller, more deeply cut leaves, smaller sulphur-yellow heads with fewer flowers, the outer involucral bracts spreading or partly erect, not reflexed, the akenes bright red or reddish brown and the pappus a purer white. *M**, among rocks in woods, Stoneham: — *B B**, occasional on rocks near The Falls.

LACTUCA, Tourn. LETTUCE.

L. **Canadensis**, L. WILD LETTUCE.
Rich soil, roadsides, clearings, and open woods. *B* and *M*, common: — *S*, occurs.

L. **integrifolia**, Bigel.
Open woods. *S**, rare; low ground.

L. **hirsuta**, Muhl.
Dry open woods. *B**, occasional; by pool in Pine Tree Brook, etc.: — *M**, Virginia Wood, etc.

SONCHUS, L. SOW THISTLE.

S. OLERACEUS, L. COMMON SOW THISTLE.
Nat. from Eu. in yards, gardens, and waste grounds. *B**, house-site.

S. ASPER, Vill. SPINY-LEAVED SOW THISTLE.
Nat. from Eu. in yards, gardens, and waste places. *M*, abundant in field, W. of Bear Hill.

† See Bot. Gaz. xx. 323.

LOBELIACEÆ. Lobelia Family.

LOBELIA, L.

L. cardinalis, L. Cardinal Flower.
Borders of brooks, and swamps. *B**, occasional; by Old Furnace Brook, etc. :— *M*, frequent in wet shade : —*S*, bog.

L. spicata, Lam.
Moist open fields and damp places. *B**, occurs :— *M*, frequent in meadows and damp borders of Winchester Reservoirs.

L. inflata, L. Indian Tobacco.
Dry or moist open fields and roadsides. *B** and *M**, common : — *S**, road to Turtle Pond.

L. Dortmanna, L. Water Lobelia.
Borders of ponds, often in shallow water. *B**, rare; by Hoosicwhisick Pond.

CAMPANULACEÆ. Campanula Family.

SPECULARIA, Heist. Venus's Looking-glass.

S. perfoliata, A. DC.
Sterile rocky places. *B*, frequent; Hancock Hill; Rattlesnake Hill, etc. : — *M**, frequent on rocks in open woods.

CAMPANULA, Tourn. Bellflower.

C. aparinoides, Pursh. Marsh Bellflower.
Wet grassy meadows and brooksides. *B*, occurs : — *M*, occasional.

C. divaricata, Michx.
Native South. *B*, old garden, Park's place, Hawk Hill: — *S*, escaped on Washington St.

C. rapunculoides, L.
Cult. in old gardens from Eu. and sparingly nat. *B*, housesite on Hillside St.; Great Blue Hill.

ERICACEÆ. Heath Family.

GAYLUSSACIA, HBK. Huckleberry.

G. dumosa, Torr. & Gray. Bog Huckleberry.
Swamps and sphagnous bogs. *S**, rare; border of Turtle Pond.

G. frondosa, Torr. & Gray. Dangleberry.
Moist open thickets. *B*, frequent; dry slopes and edges of wet open ground and in thin shade; sprouts freely from the root when cut or burnt off: — *M**, occasional: — *S**, Rooney's Rock.

G. resinosa, Torr. & Gray. Common Huckleberry.
Dry fields and open woods; recovers quickly after fires. *B**, *M*, *S** and *BB*, common.

VACCINIUM, L. Blueberry. Cranberry.

V. Pennsylvanicum, Lam. Dwarf Blueberry.
Dry hills and open pastures; in dense shade it soon disappears. *B**, *M*, *S* and *BB*, common: — the black-fruited form occurs rarely in *B*.

V. vacillans, Soland. Low Blueberry.
Dry open places; persists in abandoned pastures. *B**, *M*, *S** and *BB*, common.

V. corymbosum, L. High-bush Blueberry.
Low woods, meadows, borders of ponds, and roadsides; not persisting in heavy shade; easily killed by fire, but sprouting from the old stumps. *B* and *M*, common: — *BB**, by lower pond: — *M** and *S**, forms approaching var. *amœnum*, Gray.

Var. **amœnum,** Gray.
Same habitat as the species. *B*, by Old Furnace Brook.

Var. **atrococcum,** Gray.
Same habitat as the species. *M**, common.

V. Oxycoccus, L. Small Cranberry.
Bogs. *S**, rare.

V. macrocarpon, Ait. LARGE CRANBERRY.

Bogs and wet places. *B**, frequent : — *M*, common in wet meadows : — *S **, bog by Turtle Pond.

ARCTOSTAPHYLOS, Adans. BEARBERRY.

A. Uva-ursi, Spreng.

Bare hills, rocks, and sandy places; an important ground-cover for hill-tops, not often injured by fire. *B **, occasional; summit of Great Blue Hill, Hancock Hill, etc. : — *M*, rare : — *S*, abundant on Bearberry Hill.

EPIGÆA, L. TRAILING ARBUTUS.

E. repens, L. MAYFLOWER. TRAILING ARBUTUS.

Sandy or rocky soil, generally in light shade. *B*, very rare; a small patch only.

GAULTHERIA, Kalm. AROMATIC WINTERGREEN.

G. procumbens, L. CREEPING WINTERGREEN. CHECKERBERRY.

Damp woods and pastures, generally in shade. *B ** and *M*, common.

ANDROMEDA, L.

A. ligustrina, Muhl.

Wet and dry places in open or shady ground. *B**, common : — *M**, frequent : — *S* and *B B*, occurs.

LEUCOTHOE, Don.

L. racemosa, Gray.

Moist woods. *B **, occasional; usually about pools and on the edge of wet land; Sawcut Notch; valley, E. of Shadow Point, etc. : — *M **, frequent in low ground on borders of woods, and on S. and W. shores of Spot Pond : — *S*, meadow, N. W. of Bold Knob.

CASSANDRA, Don. LEATHER LEAF.

C. calyculata, Don.

Bogs and marshy places. *B* and *S **, common : — *M*, frequent.

KALMIA, L. AMERICAN LAUREL.

K. latifolia, L. MOUNTAIN LAUREL.

Moist rocky woods and damp places. $B*$, a considerable quantity in a single place; it must be carefully preserved.

K. angustifolia, L. SHEEP LAUREL. LAMBKILL.

Open pastures and hillsides. $B*$, common: — M, common in open places in woods and on borders of meadows: — S, frequent: — BB, occasional.

RHODODENDRON, L. ROSE BAY. AZALEA.

R. viscosum, Torr. CLAMMY AZALEA. WHITE SWAMP HONEYSUCKLE. SWAMP AZALEA.

Bogs and swampy places, in sun or light shade. $B*$, frequent: — M, common: — $S*$, occasional.

Var. glaucum, Gray.

Same general habitat as the species. $B*$, occasional; E. of Great Dome, etc.: — M, occasional: — $S*$, border of Turtle Pond, etc.

R. Rhodora, Don. RHODORA.

Swamps and bogs. B, rare: — M, occasional.

CLETHRA, Gronov. WHITE ALDER.

C. alnifolia, L. SWEET PEPPERBUSH. WHITE ALDER.

Swamps and wet places; fire kills the tops, but new stems spring from the roots with a thick and luxuriant foliage. $B*$, common about all pools, in all bogs and in shady wet runs and woods: — M, common on borders of swamps and ponds: — $S*$, border of Turtle Pond.

CHIMAPHILA, Pursh. WINTERGREEN.

C. umbellata, Nutt. PRINCE'S PINE. PIPSISSEWA. WINTERGREEN.

Dry woods. $B*$, common: — $M*$ and $S*$, frequent.

C. maculata, Pursh. SPOTTED WINTERGREEN.

Dry woods. $B*$, occasional; Hancock Hill, etc.: — M, rare.

MONESES, Salisb.   ONE-FLOWERED PYROLA.

M. grandiflora, Salisb.
Dry shady woods. *M*, occasional on Bear Hill.

PYROLA, Tourn.   PYROLA.   SHINLEAF.

P. chlorantha, Swz.
Open woods. *M*, rare.

P. elliptica, Nutt.   SHINLEAF.
Rich, moist or dry woods. *B** and *M**, common.

P. rotundifolia, L.   ROUND-LEAVED PYROLA.
Rich, moist or dry woods. *B*, common: — *M**, frequent: — *S**, occasional.

MONOTROPA, L.   INDIAN PIPE.   PINESAP.

M. uniflora, L.   INDIAN PIPE.
Rich and dry woods. *B**, common: — *M*, frequent: — *S**, occurs.

M. Hypopitys.   PINESAP.
Dry woods. *B**, occasional; Hancock Hill, etc.: — *M**, frequent.

PRIMULACEÆ.   PRIMROSE FAMILY.

HOTTONIA, L.   FEATHERFOIL.   WATER VIOLET.

H. inflata, Ell.
In still water. *M**, occasional in stagnant water, ditches, and bogs: — *S**, rare.

TRIENTALIS, L.   STAR-FLOWER.

T. Americana, Pursh.   STAR-FLOWER.   STAR ANEMONE.
Damp or dry woods. *B* and *M**, common.

STEIRONEMA, Raf.

S. ciliatum, Raf.
Low grounds and thickets. *M**, rare.

S. lanceolatum, Gray.
Low grounds and woods. *B**, *S** and *B B*, rare: — *M*, occasional.

LYSIMACHIA, Tourn. LOOSESTRIFE.

L. VULGARIS, L.
Occasionally nat. from Eu. *B*, old garden, Park's place, Hawk Hill.

L. quadrifolia, L. WHORLED LOOSESTRIFE.
Open woods, low ground, and roadsides. *B** and *M*, common: — *S**, frequent.

L. stricta, Ait. SWAMP LOOSESTRIFE.
Swamps and low ground. *B* and *S**, frequent: —*M*, common.

L. NUMMULARIA, L. MONEYWORT.
Nat. from Eu. and escaped from gardens. *M*, common in field near Spot Pond.

L. thyrsiflora, L. TUFTED LOOSESTRIFE.
Swamps and bogs. *M*, rare; swamp near Bear Hill.

OLEACEÆ. OLIVE FAMILY.

FRAXINUS, Tourn. ASH.

F. Americana, L. WHITE ASH.
Moist woods and open places. *B*, common; it seems to prefer the rich moist land by running water, though it is found almost anywhere; Marigold Brook valley; Sassamon Notch; by Monatiquot Stream, etc. :— *M*, common in all situations excepting very wet ground; S. of North Dam, a tree 5 ft. 7 in. in circumference : — *S **, occasional : — *B B*, numerous young trees in both sections; several fine trees near N. W. corner of S. section.

F. pubescens, Lam. RED ASH.
Swamps and low ground. *B*, occasional, sometimes even on rocky hillsides : — *M*, rare; Forest St. : — *S*, swamp near Turtle Pond.

F. sambucifolia, Lam. BLACK ASH.
Swamps and damp woods. *B*, occasional ; Blueberry Swamp; near Balster Brook, etc. :— *M*, rare ; a few small trees scattered in low lands : — *S*, near Turtle Pond.

SYRINGA, L. LILAC.

S. VULGARIS, L.

Cult. from E. Eu. and escaped. *B*, old garden, Canton Ave. : — *M*, occasional.

LIGUSTRUM, Tourn. PRIVET.

L. VULGARE, L.

Cult. for hedges from E. Eu. and run wild. *B**, frequent; generally confined to localities where there is evidence of a house-site; Pine Tree Brook valley; rocks near summit of Great Blue Hill, etc. : — *M*, frequent by roadsides and on rocky hills; abundant on Bear Hill : — *S*, common; house-sites, pastures, and ledges : — *BB*, occasional.

APOCYNACEÆ. DOGBANE FAMILY.

APOCYNUM, Tourn. DOGBANE.

A. androsæmifolium, L. SPREADING DOGBANE.

Open woods, fields, and roadsides. *B**, *M** and *S**, common.

A. cannabinum, L. INDIAN HEMP.

Moist ground. *M**, rare; noted but once.

ASCLEPIADACEÆ. MILKWEED FAMILY.

ASCLEPIAS, L. MILKWEED.

A. incarnata, L., var. **pulchra,** Pers. SWAMP MILKWEED.

Swamps and wet places. *B**, occasional : — *M**, common in meadows and by borders of ponds : — *S**, occasional.

A. Cornuti, Decaisne. COMMON MILKWEED.

Open fields and roadsides, a most pernicious weed, spreading rapidly from its deep, thick perennial rootstocks. *B* and *M*, common.

A. phytolaccoides, Pursh. POKE MILKWEED.

Open woods and roadsides. *B**, frequent; Randolph Ave. ;

Hawk Hill, etc.: — *M*, rare; by Spot Pond: — *S*, rare in damp woods.

A. quadrifolia, L.
Dry shady woods. *B**, occasional in a few localities: — *M*, generally distributed, but not common: — *S**, near Rooney's Rock.

A. verticillata, L.
Dry soil and rocky ledges. *S**, rare.

GENTIANACEÆ. Gentian Family.

BARTONIA, Muhl.

B. tenella, Muhl.
Damp woods. *B**, occasional; Cedar Swamp; Pine Tree Brook valley, etc.

MENYANTHES, Tourn. Buckbean.

M. trifoliata, L.
Bogs. *B* and *S**, rare.

LIMNANTHEMUM, Gmelin. Floating Heart.

L. lacunosum, Griseb.
Shallow water. *B*, rare: — *S**, Turtle Pond.

BORRAGINACEÆ. Borrage Family.

CYNOGLOSSUM, Tourn. Hound's Tongue.

C. officinale, L. Common Hound's Tongue.
Nat. from Eu., a coarse weed in pastures, yards, and by roadsides. *B*, rare; Randolph Ave.: — *M*, rare; roadside by Spot Pond.

MYOSOTIS, Dill. Forget-me-not.

M. palustris, Withering. True Forget-me-not.
Nat. from Eu. in wet ground. *BB**, rare.

M. laxa, Lehm.

Meadows, streams, and ditches. *B**, occurs: —*M*, common in bogs and wet grounds: —*S** and *BB**, occasional.

M. verna, Nutt.

Dry thin soil. *M**, frequent by roadsides and in dry rocky woods: —*BB**, frequent; ledge by lower pond.

ECHIUM, Tourn. VIPER'S BUGLOSS.

E. VULGARE, L. BLUE WEED.

Nat. from Eu. by roadsides and in meadows and waste places. *M*, rare.

CONVOLVULACEÆ. CONVOLVULUS FAMILY.

IPOMŒA, L. MORNING GLORY.

I. HEDERACEA, Jacq.

Cult. from Trop. Amer. and run wild. *B*, rare.

I. PURPUREA, Lam. COMMON MORNING GLORY.

Cult. from Trop. Amer. and run wild. *B**, house-site, Hillside St.

CONVOLVULUS, Tourn. BINDWEED.

C. sepium, L. WILD MORNING GLORY. HEDGE BINDWEED.

Low grounds, also planted. *B*, spreading from old garden, Park's place, Hawk Hill.

CUSCUTA, Tourn. DODDER.

C. Gronovii, Willd.

Moist thickets and wet places, spreading over herbs and low bushes and becoming entirely parasitic. *B** and *M*, occasional: —*S**, occurs.

SOLANACEÆ. NIGHTSHADE FAMILY.

LYCOPERSICUM, Mill. TOMATO. LOVE APPLE.

L. ESCULENTUM, Mill. TOMATO.

Cult. from Trop. Amer. *B**, escape by road.

SOLANUM, Tourn. NIGHTSHADE.

S. DULCAMARA, L. BITTERSWEET.

Nat. from Eu. iu moist places. *B**, common among loose rocks at the base of ledges, by streams, etc. : —*M*, common on borders of meadows and ponds and in low woods : —*S ** and *B B**, occasional.

PETUNIA, Juss.

P. NYCTAGINIFLORA, Juss.

Cult. from S. Amer. *B**, old gardens.

SCROPHULARIACEÆ. FIGWORT FAMILY.

VERBASCUM, L. MULLEIN.

V. THAPSUS, L. COMMON MULLEIN.

Nat. from Eu. in fields, open woods, and by roadsides. *B* and *M*, common.

LINARIA, Tourn. TOAD FLAX.

L. Canadensis, Dumont. WILD TOAD FLAX.

Sandy open places. *B**, frequent : — *M*, common along paths and roads in dry woods : — *S**, common.

L. VULGARIS, Mill. BUTTER-AND-EGGS.

Nat. from Eu. in fields and waste places. *B**, rare : — *M*, common : — *S* and *B B*, occasional.

CHELONE, Tourn. SNAKE-HEAD. TURTLE-HEAD.

C. glabra, L.

Wet places. *B*, frequent : — *M*, common along streams and ponds : — *B B*, occasional.

MIMULUS, L. MONKEY FLOWER.

M. ringens, L.

Wet shady places, meadows and swamps. *B**, occasional ; Pine Tree Brook valley, etc. : — *M*, common.

GRATIOLA, L. Hedge Hyssop.

G. aurea, Muhl.

Wet sandy or gravelly places. *B* *, rare : — *M* *, abundant in wet meadows and on borders of Spot Pond, etc.

ILYSANTHES, Raf.

I. riparia, Raf. False Pimpernel.

Wet gravelly or sandy shores. *B* *, frequent ; by pool in Pine Tree Brook, etc. : — *M* *, rare.

VERONICA, L. Speedwell.

V. scutellata, L. Marsh Speedwell.

Bogs and ditches. *B* * and *M*, common : — *S* *, wet meadows : — *B B* *, frequent.

V. officinalis, L. Common Speedwell.

Open dry or moist woods. *M* *, rare ; a large clump in grassy place in damp woods.

V. serpyllifolia, L. Thyme-leaved Speedwell.

Fields, banks, and open sunny places. *B* *, occasional : — *M*, common in paths and open woody places : —*B B* *, frequent.

V. peregrina, L. Purslane Speedwell.

Cult. and waste ground. *B B*, rare ; by the brook.

V. arvensis, L. Corn Speedwell.

Nat. from Eu. in cult. ground and waste places. *M* *, rocky woods on slope of Bear Hill : — *B B*, rare ; by the brook.

GERARDIA, L.

G. pedicularia, L.

Dry open woods. *B* *, common : — *M* *, frequent : — *S* *, occurs.

G. flava, L. Downy False Foxglove.

Dry open woods. *B* *, common : — *M* *, frequent on rocky wooded hillsides : — *S* *, N. of Turtle Pond, etc.

G. quercifolia, Pursh. Smooth False Foxglove.

Dry open woods. *B* *, frequent ; woods on slope of Great Blue Hill ; top of Buck Hill, etc.

G. purpurea, L., var. paupercula, Gray. PURPLE GE-
RARDIA.
Low, partly open grounds. *B* * and *M* *, common.

G. tenuifolia, Vahl. SLENDER GERARDIA.
Woods and partly open ground. *B* * and *M*, common: —
S *, occasional.

PEDICULARIS, Tourn. LOUSEWORT.

P. Canadensis, L. COMMON LOUSEWORT. WOOD BETONY.
Woods and moist places. *B* * and *M*, frequent.

MELAMPYRUM, Tourn. COW WHEAT.

M. Americanum, Michx.
Dry open woods. *B* *, *M* and *S* *, common: — *B B*, occasional.

OROBANCHACEÆ. BROOM-RAPE FAMILY.

EPIPHEGUS, Nutt. BEECHDROPS.

E. Virginiana, Bart.
Under beech trees, parasitic on their roots. *B* *, frequent:
—*M*, rare.

CONOPHOLIS, WALLROTH. SQUAW-ROOT. CANCER-ROOT.

C. Americana, WALLROTH.
Oak woods among decaying leaves. *M* *, very rare; one patch in the E. part, and one in the S. W. part of the Reservation.

APHYLLON, Mitchell. NAKED BROOM-RAPE.

A. uniflorum, Gray. ONE-FLOWERED BROOM-RAPE.
Low ground and grassy places. *B* *, rare; Rattlesnake Hill: — *M* *, occasional: — *S* *, rare; slope of Bellevue Hill.

LENTIBULARIACEÆ. BLADDERWORT FAMILY.

UTRICULARIA, L. BLADDERWORT.

U. vulgaris, L. GREATER BLADDERWORT.

Ponds and slow streams. *M*, ditch in meadow near Spot Pond.

U. gibba, L.
Shallow water, and muddy and sandy places. *M*, Spot Pond.

U. intermedia, Hayne.
Bogs and streams. *B*, rare; pools.

U. purpurea, Walt. LARGE PURPLE BLADDERWORT.
Shallow water and streams. *M*, Spot Pond.

U. cornuta, Michx. LONG-SPURRED BLADDERWORT.
Bogs and sandy shores. *B*, rare; by Monatiquot Stream: — *S**, bog by Turtle Pond.

BIGNONIACEÆ. BIGNONIA FAMILY.

CATALPA, Scop. CATALPA. INDIAN BEAN.

C. bignonioides, Walt.
Cult. from Ga., Ala., and Miss. *B*, old garden, Park's place, Hawk Hill.

VERBENACEÆ. VERVAIN FAMILY.

VERBENA, Tourn. VERVAIN.

V. urticæfolia, L. WHITE VERVAIN.
Meadows, and dry or moist open places. *B*, occasional: — *M*, frequent by roadsides and in open woody places.

V. hastata, L. BLUE VERVAIN.
Meadows and dry or moist open places. *M*, occasional.

LABIATÆ. MINT FAMILY.

TRICHOSTEMA, L. BLUE CURLS.

T. dichotomum, L. BASTARD PENNYROYAL.
Sandy fields and gravelly places. *B**, *M* and *S*, common: — *BB**, occurs.

MENTHA, Tourn. MINT.

M. Canadensis, L. WILD MINT.
Meadows and roadsides. *B**, rare: — *M**, common; borders of wet grounds and Reservoirs.

LYCOPUS, Tourn. WATER HOREHOUND.

L. Virginicus, L. BUGLEWEED.
Swamps, meadows, and low ground. *B*, rare ; off Randolph Ave. : —*M**, common in bogs and swamps : —*S*, low ground.

L. sinuatus, Ell.
Swamps, meadows, and wet places. *B** and *M*, common.

PYCNANTHEMUM, Michx. MOUNTAIN MINT. BASIL.

P. linifolium, Pursh.
Dry open woods and hillsides. *B**, common on Great Blue Hill.

P. muticum, Pers.
Open woods and roadsides. *B*, occurs.

P. clinopodioides, Gray.
Dry open woods. *B*, rare.

P. incanum, Michx.
Dry open woods. *B**, occasional; slopes of Great Blue Hill, Chickatawbut Hill, Hawk Hill, etc.

THYMUS, Tourn. THYME.

T. SERPYLLUM, L. CREEPING THYME.
Cult. and occasionally run wild in old fields and waste places. *B*, rare ; a single locality.

HEDEOMA, Pers. MOCK PENNYROYAL.

H. pulegioides, Pers. AMERICAN PENNYROYAL.
Dry fields and open sunny places. *B** and *M*, common.

NEPETA, L. CATNIP.

N. CATARIA, L. CATNIP.
Nat. from Eu. about houses and roadsides. *B**, common : —*M*, roadsides, etc.

N. GLECHOMA, Benth. GROUND IVY. GILL-OVER-THE-GROUND.
Nat. from Eu. in cult. and waste ground, and on roadsides. *B**, occasional : —*M*, roadside, Forest St.

SCUTELLARIA, L. SKULLCAP.

S. lateriflora, L. MAD-DOG SKULLCAP.
Wet shady places. *B**, frequent: —*M**, frequent in bogs and swamps.

BRUNELLA, Tourn. SELF-HEAL.

B. vulgaris, L. COMMON SELF-HEAL.
Open woods, fields, and roadsides. *B**, *M**, *S** and *BB*, common.

LEONURUS, L. MOTHERWORT.

L. CARDIACA, L. COMMON MOTHERWORT.
Nat. from Eu. about dwellings and in waste places. *B**, *M*, *S** and *BB*, occasional.

GALEOPSIS, L. HEMP NETTLE.

G. TETRAHIT, L. COMMON HEMP NETTLE.
Nat. from Eu. in all waste places. *B**, rare; house-sites.

STACHYS, Tourn. HEDGE NETTLE.

S. aspera, Michx.
Wet places. *B**, rare.

PLANTAGINACEÆ. PLANTAIN FAMILY.

PLANTAGO, Tourn. PLANTAIN.

P. major, L. COMMON PLANTAIN.
Roadsides, fields, paths, waste places, and thin woods. *B**, *M* and *BB*, common:— *S*, frequent.

P. Rugelii, Decaisne.
Similar situations with the last. *B*, frequent:— *M**, common; Cascade woods; by S. Reservoir, etc.

P. LANCEOLATA, L. RIBGRASS.
Nat. from Eu. in fields, open woods, roadsides, and waste places everywhere. *B**, *M*, *S* and *BB**, common.

P. PATAGONICA, Jacq., var. ARISTATA, Gray.
Nat. from the West by roadsides, etc. *B**, rare.

DIVISION III. APETALÆ.

ILLECEBRACEÆ. Knotwort Family.

ANYCHIA, Michx. Forked Chickweed.

A. dichotoma, Michx.
Dry open woods. $B*$, occasional.

A. capillacea, DC.
Dry open woods. $B*$, occasional : — $M*$, rare.

SCLERANTHUS, L. Knawel.

S. annuus, L.
Nat. from Eu. in waste places. M, common : —$B_l B*$, occurs.

AMARANTACEÆ. Amaranth Family.

AMARANTUS, Tourn. Amaranth.

A. retroflexus, L. Amaranth Pigweed.
Adv. from Trop. Amer. in cult. ground, by roadsides, and in all waste places. $B*$, top of Great Blue Hill : — M, occasional.

CHENOPODIACEÆ. Goosefoot Family.

CHENOPODIUM, Tourn. Pigweed.

C. album, L. Pigweed.
Nat. from Eu. in cult. grounds and waste places everywhere; a most troublesome weed. $B*$, roadside, E. of Great Dome : — $M*$, roadsides and old cult. ground.

C. murale, L.
Adv. from Eu. in waste places. $B*$, old cult. field, W. Quincy.

C. hybridum, L. Maple-leaved Goosefoot.
Int. from the West into cult. grounds, waste places, and roadsides. M, rare ; roadside.

C. ambrosioides, L. Mexican Tea.
Nat. from Trop. Amer. in waste places. $B*$, rare.

PHYTOLACCACEÆ. Pokeweed Family.

PHYTOLACCA, Tourn. Pokeweed.

P. decandra, L. Common Poke. Garget. Pigeon Berry.

Low grounds, rocky slopes, and open woods; conspicuously vigorous after fires. *B** and *M*, occasional.

POLYGONACEÆ. Buckwheat Family.

RUMEX, L. Dock. Sorrel.

R. crispus, L. Curled Dock.

Nat. from Eu. in cult. and waste ground. *B*, frequent: — *M*, common: — *S**, occasional.

R. obtusifolius, L. Bitter Dock.

Nat. from Eu. by roadsides, and in all waste places. *B*, occasional; Hillside St., etc.: — *M**, frequent in bogs and by low roadsides: — *S**, rare.

R. Acetosella, L. Field Sorrel.

Nat. from Eu. in fields, by roadsides, and in all waste places. *B*, *M*, *S** and *B B**, common.

R. Acetosa, L. Sorrel Dock.

Sparingly nat. from Eu. in waste places. *M*, a single plant in moist ground, N. E. of Bear Hill.

POLYGONUM, Tourn. Knotweed.

P. aviculare, L. Doorweed.

Yards, waste places, etc. *B**, top of Great Blue Hill, by Observatory.

P. tenue, Michx.

Dry open places. *B**, frequent on hill-tops.

P. lapathifolium, L., var. **incarnatum,** Watson.

Wet places. *M**, abundant on W. shore of Spot Pond, showing a gradation into the var. *incanum*, Koch.

Var. **incanum,** Koch.

Wet places. *M**, abundant on the stony S. and W. shores of Spot Pond, fruiting from one to six inches high.

P. Muhlenbergii, Watson.

Bogs and wet places. B^*, rare; valley, E. of Shadow Point; pool near Great Dome.

P. Careyi, Olney.

Swamps and shady wet places. B, rare; near Balster Brook.

P. Persicaria, L. Lady's Thumb.

Nat. from Eu. in cult. ground and damp waste places. B^* and M, common.

P. hydropiperoides, Michx. Mild Water Pepper.

Shallow water and wet places. B^*, pool, E. of Great Dome, etc.: — M^*, frequent in ditches and low grounds: — BB^*, border of pond.

P. Hydropiper, L. Common Smartweed. Water Pepper.

Wet places. B^*, frequent; pool, E. of Great Dome; by Cragfoot Spring, etc.: — M^* and S, common.

P. acre, HBK. Water Smartweed.

Wet places. B^*, occasional; pool in Pine Tree Brook; Old Furnace Brook; by Cragfoot Spring, etc.: — M, by Winchester Reservoirs: — BB^*, occurs.

P. Virginianum, L.

Damp woods. S, rare.

P. arifolium, L. Halberd-leaved Tear-thumb.

Bogs in shade. B, occasional: — M^* and S, frequent.

P. sagittatum, L. Arrow-leaved Tear-thumb.

Low ground. B^* and S, common: —M, common in wooded bogs and swamps.

P. Convolvulus, L. Black Bindweed.

Nat. from Eu. in cult. and waste grounds. B, occasional: —M, borders of cult. ground.

P. dumetorum, L., var. **scandens**, Gray. Climbing False Buckwheat.

Moist woods. B^*, occasional; near Cragfoot Spring, etc. —M, occasional; by S. Reservoir, etc.

FAGOPYRUM, Tourn. BUCKWHEAT.

F. ESCULENTUM, Moench.
Cult. from N. Asia, and occasionally escaped. *M*, near Winchester Reservoirs.

LAURACEÆ. LAUREL FAMILY.

SASSAFRAS, Nees.

S. officinale, Nees.
Dry or moist woods, and open ground. *B*, *M* and *S* *, frequent; dry slopes and hill-tops throughout; easily killed by fires, but sprouts freely.

LINDERA, Thunb. FEVER BUSH. WILD ALLSPICE.

L. Benzoin, Blume. SPICE BUSH. BENJAMIN BUSH.
Rich damp woods; prefers shade and wet but not stagnant soil and is generally found in shady wet runs; very sensitive to fire, and only young plants will recover readily when injured. *B**, *M**, *S* and *B B**, frequent.

SANTALACEÆ. SANDALWOOD FAMILY.

COMANDRA, Nutt. BASTARD TOAD FLAX.

C. umbellata, Nutt.
Thin dry soil, in open ground and rocky woods. *B* * and *M*, common : — *S* *, occasional.

EUPHORBIACEÆ. SPURGE FAMILY.

EUPHORBIA, L. SPURGE.

E. maculata, L.
Dry open ground. *M**, gravelly shore of N. Reservoir.

E. CYPARISSIAS, L. GRAVEYARD FLOWER. CYPRESS.
Nat. from Eu. in waste places. *M**, frequent by roadsides and old houses.

ACALYPHA, L. THREE-SEEDED MERCURY.

A. Virginica, L.
Poor soil in sun or shade. *B**, *M* and *S*, frequent.

Var. **gracilens**, Muell.
Sandy soil. *M**, shore of N. Reservoir.

URTICACEÆ. Nettle Family.

ULMUS, L. Elm.

U. Americana, L. American Elm.

Moist woods, rich soil by river banks, etc. *B*, planted trees are frequent by roadsides and old house-sites; natives are rare and are usually found in wet places; one tree in a group near the head of Marigold Brook valley has a diameter of 5 ft. 3 in., and spread of 50 ft.; small trees at head of Beech Run: — *M*, common in pastures, open moist places, and occasionally in woods; planted by roadsides: —*S*, moist valley, W. of Turtle Pond; planted on Washington St.: —*BB*, numerous trees in both sections; in S. section near the road is the largest elm in all the Reservations, its circumference, 4 ft. above the ground, being 17 ft. 5 in.

MORUS, Tourn. Mulberry.

M. alba, L. White Mulberry.
Cult. from Eu. *B*, old garden, Park's place, Hawk Hill.

URTICA, Tourn. Nettle.

U. gracilis, Ait.
Moist ground in waste places and by roadsides. *M*, common.

PILEA, Lindl. Richweed. Clearweed.

P. pumila, Gray.
Moist shady places. *B**, *M* and *BB**, rare.

BŒHMERIA, Jacq. False Nettle.

B. cylindrica, Willd.
Wet shady places. *B**, rare; Cedar Swamp; by Pine Tree Brook: — *M*, rare; by stream in Virginia Wood, etc.: — *S**, rare; swamp near office: —*BB*, rare; by the brook.

PLATANACEÆ. Plane-tree Family.

PLATANUS, L. Sycamore. Buttonwood.

P. occidentalis, L.
Rich moist soil, generally along streams. *B*, occasional small trees; by Pine Tree Brook; Beech Run, etc. : — *M*, occasional by roadsides and in moist land; large trees by swamp on S. E. side of Bear Hill: — *B B*, frequent near running water.

JUGLANDACEÆ. Walnut Family.

JUGLANS, L. Walnut.

J. cinerea, L. Butternut.
Rich woods. *B**, frequent in rich soil about the base of the hills; house-site, Hillside St. : — *M*, occasional : — *B B*, rare in the S. section; occasional in the N. section.

J. cinerea × regia, Sargent. †
B, a most interesting hybrid on the Old Houghton Place, near Hoosicwhisick Pond; it has a tall straight trunk with a diameter of 3 ft. 1 in. at 3 ft. from the ground.

CARYA, Nutt. Hickory.

C. alba, Nutt. Shell-bark Hickory. Shag-bark Hickory.
Rich soil on hillsides and in woods. *B**, frequently found in woods with oaks and chestnuts; a group by Marigold Brook valley, one tree 7 ft. in circumference with a spread of 40 ft.; valley, S. of Rattlesnake Hill, etc. : — *M*, common by roadsides and in woods.

C. tomentosa, Nutt. Mocker-Nut. White-heart Hickory.
Rich hillsides. *B*, frequent; most abundant about hill-tops where it is twisted and scrubby.

C. porcina, Nutt. Pig-Nut.
Dry uplands and hills. *B**, common; found with the Mocker-Nut near hill-tops, and also among oaks and chestnuts

† See Garden and Forest, vii. 434, f. 69.

in mixed woods: — *M**, frequent as a small tree or shrub on rocky wooded hillsides.

C. amara, Nutt. BITTER-NUT. SWAMP HICKORY.

Moist soil, borders of streams and swamps. *M*, a few trees near Pine Hill.

MYRICACEÆ. SWEET GALE FAMILY.

MYRICA, L. BAYBERRY.

M. Gale, L. SWEET GALE.

Wet borders of ponds, meadows, and edges of ditches; will not persist long in shade. *B** and *M*, occasional.

M. cerifera, L. BAYBERRY.

Sandy soil, open fields and meadows. *B**, frequent; grows naturally on hill-tops and in dry open ground on the edges of meadows, but is often crowded into wet ground by other shrubs; is soon killed out by shade: — *M*, common in pastures, and in open places in dry woods: — *S* and *BB*, occasional.

M. asplenifolia, Endl. SWEET FERN.

Sterile soil in fields and open places; an important ground-covering shrub in burnt land, quickly covering the surface where the growth of trees is thin; recovers quickly if killed by fire, but will not grow in dense shade. *B**, *M**, *S** and *BB*, common.

CUPULIFERÆ. OAK FAMILY.

BETULA, Tourn. BIRCH.

B. lenta, L. BLACK BIRCH. CHERRY BIRCH. SWEET BIRCH.

Rich woodlands. *B*, occasional; prefers cool constantly moist soil and will grow in wet ground; near summit of Great Blue Hill; valley, S. of Rattlesnake Hill; near Crossman Pines, trees 40 to 50 ft. high; near Hillside Pond is a fine large tree, about 15 in. in diameter, growing between and forcing apart two great bowlders with its trunk, which reaches just to their tops, then branches out and forms a spreading canopy over them: — M^*, occasional on dry slopes and ledges; fine trees along Owen's and Flagg's Walks; slope of Bear Hill, etc. : — S^*, N. W. of Bold Knob.

B. lutea, Michx. f. YELLOW BIRCH.

Rich moist woods. B^*, occasional; Eagle Valley; glen above Crossman Pines; Bear Hill: — M, occasional; large trees near Cascade; Flagg's Walk; fine trees by a brook, N. of Pine Hill, near Forest St.

B. populifolia, Ait. WHITE BIRCH. GRAY BIRCH.

Dry barren soil, and borders of swamps. B and M^*, common everywhere in open ground from the bleak hill-top to the wet swamp; young trees are easily killed by fire, but sprout readily; it is soon shaded out by overtopping trees: — S, occasional in open ground; less frequent in swamps: — BB, frequent in wet ground and edges of pastures.

B. papyrifera, Marsh. PAPER BIRCH. CANOE BIRCH.

Rich woods and banks of streams. M, rare; edge of pool, N. W. of Cascade; one tree, W. edge of Cranberry Pool.

ALNUS, Tourn. ALDER.

A. incana, Willd. SPECKLED ALDER. HOARY ALDER.

Borders of streams and swamps. B, frequent; more common about the meadows and bogs, at a low station, than higher up the slopes; will not persist long in shade: — S and BB^*, common.

A. serrulata, Willd. SMOOTH ALDER.

Borders of streams and swamps. B^*, common: — M^*, common on borders of ponds and meadows; along Pine Hill brook: — BB^*, by the brook.

CORYLUS, Tourn. HAZEL-NUT.

C. Americana, Walt. WILD HAZEL-NUT.

Open thickets and roadsides. B^*, frequent; usually found in dry places in open land; recovers quickly after a fire, but spreads slowly; will not persist permanently in shade; W. of

Pasture Run; Pine Tree Brook valley, etc. : — *M* and *S*, common in same habitat as above : — *B B*, occasional.

C. rostrata, Ait. BEAKED HAZEL-NUT.

Open thickets and roadsides. *B* *, occasional ; the common species along the northerly line of the Reservation ; base of Hawk Hill ; W. side of Cedar Swamp ; W. slope of Great Dome : — *M*, occasional ; abundant along S. E. slopes of S. Reservoir.

OSTRYA, Mich. HOP HORNBEAM. IRON WOOD.

O. Virginica, Willd. AMERICAN HOP HORNBEAM. LEVER WOOD.

Dry woods and hillsides. *B* *, common ; broken ledgy slopes, and dry gravelly woods ; grows well in shade : —*M*, common ; large tree between spring and summit of Silver Mine Hill : —*S*, occasional.

CARPINUS, L. HORNBEAM. IRON WOOD.

C. Caroliniana, Walt. HORNBEAM. BLUE BEECH.

Borders of streams and swamps. *B* *, occasional ; found in cool rocky ravines and on the edges of brooks ; by Pine Tree Brook ; Sassamon Notch, etc. : —*M*, rare ; roadside, S. W. of Spot Pond, etc. : —*S* *, Happy Valley.

QUERCUS, L. OAK.

Q. alba, L. WHITE OAK.

In all soils. *B* *, common from the summits of the hills to the edges of the swamps ; resists fire well and grows well in shade : —*M* and *S*, common throughout : —*B B*, common ; the famous Waverly Oaks, twenty-five in number, are of this species, with one exception, *Q. bicolor*, and they are noble trees, growing on a characteristic kame ; the Big White Oak on the N. slope of the kame is about 50 ft. high, with a circumference of 18 ft. 7½ in. at 5 ft. from the ground.

Q. bicolor, Willd. SWAMP WHITE OAK.

Borders of streams and swamps. *B* *, rare ; abundant

about Monatiquot Stream; near Streamside Ledge, etc.: —*M* and *S*, common: —*B B*, occasional in both sections; one of the large Waverly Oaks is of this species, 65 ft. high, with circumference of 12 ft. 6 in. at 5 ft. from the ground.

Q. Prinus, L. CHESTNUT OAK.

Rocky woods and hillsides. *B**, frequent in every kind of soil except very wet, and in all exposures; comparatively scarce on Great Blue Hill and in other parts of the western and middle sections, becoming more frequent in the eastern section, and in considerable numbers between Wampatuck and Rattlesnake Hills.

Q. prinoides, Willd. CHINQUAPIN OAK. DWARF CHESTNUT OAK.

Rocky woods and dry hillsides. *B**, common; usually growing with *Q. ilicifolia*, mostly on the hill-tops: —*M*, occasional.

Q. rubra, L. RED OAK.

In rich and poor soil. *B*, *M*, *S* and *B B*, common; rather less on the hill-tops and more in the wet valleys than *Q. alba*.

Q. coccinea, Wang. SCARLET OAK.

Rich ground or dry sandy soil. *B*, frequent; sometimes in considerable numbers, but oftener as scattered trees: —*M*, *S* and *B B*, occasional.

Var. **tinctoria**, Gray. BLACK OAK.

Dry hills and gravelly uplands. *B*, occasional; not as common as the type: —*M*, *S* and *B B*, frequent.

Q. ilicifolia, Wang. BEAR OAK. BLACK SCRUB OAK.

Rocky hills and sterile places. *B**, common; on all hill-tops, ledges and barren places; sprouts freely after fires; large field covered with it on N. slope of Rattlesnake Hill; predominates on tops of Babel Rock, Great Dome, Great Blue Hill, etc.: —*M*, *S** and *B B*, common in similar situations.

CASTANEA, Tourn. CHESTNUT.

C. sativa, Mill., var. **Americana**, Watson.

Rocky woods and dry hillsides. *B*, common; found throughout on well drained slopes; sprouts more freely than any other tree, old sprouts resisting fire well; large tree, N. E. side of Hoosicwhisick Pond, 17 ft. in circumference, 83 ft. high, 50 ft. spread: —*M*, small trees frequent in woods; grove of large trees in N. E. part of the Reservation.

FAGUS, Tourn. BEECH.

F. ferruginea, Ait. AMERICAN BEECH.

Rich woods. *B*, frequent; usually represented by sprouts from roots and by young trees that have persisted since older trees were cut; the trees are killed by even a slight fire, those remaining being generally protected by moist land or by ledges and rocks; Beech Run; W. slope of Blue Hill, etc.: —*M*, occasional in woods; a small grove of young trees on Owen's Walk: —*S*, valley, E. of Turtle Pond.

SALICACEÆ. WILLOW FAMILY.

SALIX, Tourn.† WILLOW.

S. nigra, Marsh. BLACK WILLOW.

Banks of streams and ponds and in wet places. *B**, rare; by Old Furnace Brook; W. of Sawcut Notch: —*S*, rare; bog in Turtle Pond woods.

S. lucida, Muhl. SHINING WILLOW.

Wet places and borders of streams and ponds. *B* and *M*, occasional: —*S **, low ground, W. of Office.

S. rostrata, Richards.

Moist or dry places. *B*, common both in the low land and on the summits of the range: —*M*, common.

S. discolor, Muhl. GLAUCOUS WILLOW. COMMON SWAMP WILLOW.

Low ground. In all the Reservations on edges of meadows, in bogs and the more open parts of swamps, and occasionally

---

† Forms of *S. alba*, L. or *S. fragilis*, L. or both occur in all the Reservations, but no reports have been received by which a determination of the species can be made.

on hill-tops, the long pistillate catkins conspicuous by the end of April:—*B** and *M*, common:—*S* and *B B*, frequent.

S. humilis × discolor, Bebb.
Same habitat as the parent species. *M**, rare; near S. Reservoir.

S. humilis, Marsh. LOW WILLOW. PRAIRIE WILLOW.
Dry or wet ground. *B**, Chestnut Run path: — *M*, frequent:— *S**, occurs.

S. tristis, Ait. DWARF GRAY WILLOW.
Sandy places, gravelly fields, low thickets and roadsides. *B**, rare; N. slope of Rattlesnake Hill; W. of Sawcut Notch: — *M**, occasional:— *S**, Happy Valley.

S. sericea, Marsh. SILKY WILLOW.
Low ground. *B**, rare; Chestnut Run path, E. of Hancock Hill: — *M**, rare; N. side of Spot Pond.

S. petiolaris, Smith.
Low ground. *M**, rare; fields round Spot Pond; meadow, N. of Bear Hill; S. end of S. Reservoir.

S. PURPUREA, L. PURPLE WILLOW.
Adv. from Eu. in low ground. *M*, one tree in low land, S. W. of Pine Hill.

S. cordata, Muhl. HEART-LEAVED WILLOW.
Wet places, along streams and by damp roadsides. *B*, frequent: — *M*, occasional.

S. cordata × sericea, Bebb.
Low ground. *B*, swamp, N. E. of Hawk Hill.

S. myrtilloides, L.
Cold bogs. *M*, occasional.

POPULUS, Tourn. POPLAR. ASPEN.

P. tremuloides, Michx. ASPEN.
Woods and roadsides; the most widely distributed tree in North America. *B*, occasional: — *M*, common.

P. grandidentata, Michx. LARGE-TOOTHED ASPEN.
Rich woods, borders of streams, and hillsides. *B**, frequent; Rattle Rock; woods near boundary, N. of Hancock Hill, etc. : — *M**, frequent as a small tree in open places in rocky woods.

P. balsamifera, L., var. CANDICANS, Gray. BALM OF GILEAD.
This tree is rare or unknown in a wild state but very common in cultivation. *B**, house-site, Hillside St. : — *M*, occasional; one large tree not far from Black Rock, another by Doleful Pond.

SUB-CLASS II. MONOCOTYLEDONES.

ORCHIDACEÆ. ORCHID FAMILY.

MICROSTYLIS, Nutt. ADDER'S MOUTH.

M. ophioglossoides, Nutt.
Low ground. *B*, very rare ; Cedar Swamp.

LIPARIS, Richard. TWAYBLADE.

L. liliifolia, Richard.
Moist woods. *B*, very rare ; base of Buck Hill.

CORALLORHIZA, Haller. CORAL-ROOT.

C. innata, R. Br.
Swamps and cool damp woods. *B*, very rare; this interesting but inconspicuous little plant grows in a single spot in the E. part of the Reservation : — *M*, rare.

C. odontorhiza, Nutt.
Rich woods. *S*, very rare ; six plants were found near Overbrook Hill in 1878.

C. multiflora, Nutt.
Dry shady woods. *B**, frequent ; Pine Tree Brook valley ; slopes of Great Blue Hill, Rattlesnake Hill, etc. : — *M*, occasional ; Virginia Wood ; N. slope of Bear Hill ; border of brook, E. of Melrose Reservoir.

SPIRANTHES, Richard. LADIES' TRESSES.

S. cernua, Richard.

Damp ground and meadows. B^*, frequent in wet grassy places; found about the top of Great Blue Hill: — M^*, common in meadows, etc.: — S, occasional.

S. gracilis, Bigel.

Dry woods and sandy places. B^*, frequent; Sawcut Notch; Pine Tree Brook valley, etc.: — M^*, frequent on rocky banks in woods.

GOODYERA, R. Br. RATTLESNAKE PLANTAIN.

G. pubescens, R. Br.

Rich woods. B^*, occasional; base of Great Blue Hill, etc.: —M, occasional.

ARETHUSA, Gronov.

A. bulbosa, L.

Bogs and wet meadows. B, rare; bog near Hoosicwhisick Pond.

CALOPOGON, R. Br.

C. pulchellus, R. Br.

Bogs and wet meadows. B, occasional; bogs about Hoosicwhisick Pond: —S^*, border of Turtle Pond.

POGONIA, Juss.

P. ophioglossoides, Nutt.

Bogs and wet meadows. B, frequent: —S, occasional; border of Turtle Pond.

P. verticillata, Nutt.

Low shady woods. B^*, occasional; slope of Great Blue Hill; N. E. of Hawk Hill, etc. This plant is very local and should be rigidly preserved.

HABENARIA, Willd.

H. tridentata, Hook.

Rich wet woods. B^*, rare; Cedar Swamp; W. of Pasture Run.

H. Hookeri, Torr.
Damp shady woods. *B*, very rare; Cedar Swamp: —*M*, rare.

H. blephariglottis, Torr. WHITE-FRINGED ORCHIS.
Bogs and wet grassy places. *B**, rare; Cedar Swamp.

H. lacera, R. Br. RAGGED-FRINGED ORCHIS.
Meadows and bogs. *B**, rare; by Hoosicwhisick Pond: — *M*, rare; by S. Reservoir.

H. psycodes, Gray.
Bogs, meadows and wet places. *B* and *M*, rare.

H. fimbriata, R. Br.
Rich woods and wet meadows. *B**, rare; one of our handsomest orchids and found in but two or three spots.

CYPRIPEDIUM, L. LADY'S SLIPPER.

C. acaule, Ait. STEMLESS LADY'S SLIPPER.
Dry open woods. *B**, frequent; Hawk Hill; N. W. slope of Fox Hill, etc.: —*M*, frequent.

IRIDACEÆ. IRIS FAMILY.

IRIS, Tourn. BLUE FLAG.

I. versicolor, L. LARGER BLUE FLAG.
Wet meadows and swamps. *B** and *M*, common: —*BB*, by lower pond.

I. GERMANICA, L. COMMON FLOWER-DE-LUCE.
Cult. from Eu. *B*, old garden, Park's place, Hawk Hill.

SISYRINCHIUM, L. BLUE-EYED GRASS.

S. anceps, Cav.
Moist grassy places. *B*, common: —*M**, frequent; border of Doleful Pond; shore of Spot Pond, etc.: —*S**, between Bellevue and Bearberry Hills: —*BB**, occurs.

AMARYLLIDACEÆ. Amaryllis Family.

HYPOXIS, L. Star Grass.

H. erecta, L.
Open woods. *B**, common: —*M**, common along wood-paths and in dry woods: —*S**, Overbrook Hill; Happy Valley, etc.: —*B B**, occurs.

LILIACEÆ. Lily Family.

SMILAX, Tourn. Greenbrier. Cat-brier.

S. herbacea, L. Carrion Flower.
Damp borders of woods and swamps. *B**, frequent; woods on Great Blue Hill, etc.: —*M*, frequent: —*S**, occurs.

S. rotundifolia, L. Common Greenbrier. Horse-brier.
Thickets. *B*, common on the edges of swamps among bushes; grows well in shade; dry place near the top of Great Blue Hill: —*M*, common in rocky woods: —*S** and *B B**, occasional by swamps and in dry soil.

S. glauca, Walt.
Dry woods. *B**, very rare; a single locality at the foot of Great Blue Hill.

ALLIUM, L. Onion. Garlic.

A. Canadense, Kalm. Wild Garlic.
Moist grassy places. *M**, numerous plants on rocks at the N. end of Doleful Pond: —*B B*, common in grass by North St.

ORNITHOGALUM, Tourn. Star-of-Bethlehem.

O. umbellatum, L.
Cult. from Eu. and nat. in fields. *B*, old garden, Park's place, Hawk Hill.

HEMEROCALLIS, L. Day Lily.

H. fulva, L. Common Day Lily.
Cult. from Old World and running wild near old houses. *B*,

old garden, Park's place, Hawk Hill. This species does not fruit in this country, while *H. flava*, L. fruits freely.

FUNKIA, Spreng.

F. LANCIFOLIA, Spreng.
Cult. from Japan. *B*, old garden, Park's place, Hawk Hill.

YUCCA, L. SPANISH BAYONET.

Y. FILAMENTOSA, L. ADAM'S NEEDLE.
Cult. from the South. *B*, old garden, Park's place, Hawk Hill.

CONVALLARIA, L. LILY OF THE VALLEY.

C. MAJALIS, L.
Cult. from Eu. and wild in the Alleghanies. *B*, old garden, Park's place, Hawk Hill.

POLYGONATUM, Tourn. SOLOMON'S SEAL.

P. **biflorum**, Ell. SMALLER SOLOMON'S SEAL.
Moist woods. *B** and *M**, common.

ASPARAGUS, Tourn.

A. OFFICINALIS, L. GARDEN ASPARAGUS.
Cult. from Eu. and frequently escaped. *B**, rare; house-site, Hillside St., etc.: — *M*, fields and roadsides; woods rather remote from cult. grounds.

SMILACINA, Desf. FALSE SOLOMON'S SEAL.

S. **racemosa**, Desf. FALSE SPIKENARD.
Rich woods and rocky wooded hillsides. *B**, *M*, *S** and *BB**, common.

MAIANTHEMUM, Wigg.

M. **Canadense**, Desf. DWARF SOLOMON'S SEAL.
Rich moist woods. *B*, *M*, *S** and *BB**, common.

UVULARIA, L. BELLWORT.

U. perfoliata, L.
Moist rich woods. *M*, very rare; Pine Hill.

OAKESIA, Watson.

O. sessilifolia, Watson. COMMON BELLWORT. WILD OATS.
Open woods. *B* and *M*, common.

ERYTHRONIUM, L. DOG-TOOTH VIOLET.

E. Americanum, Ker.
Rich shady ground. *M*, frequent on Pine Hill:—*B B**, occasional.

LILIUM, L. LILY.

L. TIGRINUM, Ker. TIGER LILY.
Cult. from China and Japan. *B*, old garden, Park's place, Hawk Hill.

L. Philadelphicum, L. WILD RED LILY. WOOD LILY.
Open places in dry woods and pastures. *B**, frequent; growing abundantly near Rattlesnake Hill; E. slope of Fox Hill, etc.:—*M*, frequent; more common in the N. part of the Reservation:—*S**, frequent.

L. Canadense, L. WILD YELLOW LILY. CANADA LILY.
Meadows and boggy places. *M*, frequent.

MEDEOLA, Gronov. INDIAN CUCUMBER ROOT.

M. Virginiana, L.
Rich woods. *B**, frequent:—*M*, common:—*S**, Rooney's Rock.

TRILLIUM, L. WAKE ROBIN.

T. cernuum, L. NODDING TRILLIUM.
Rich woods. *B**, occasional; swamp by Hawk Hill; Rattlesnake Hill, etc.:—*M**, frequent; damp woods, N. of Wright's Pond, etc.

VERATRUM, Tourn. FALSE HELLEBORE.

V. viride, Ait. INDIAN POKE.
Low grounds, and swamps. M, frequent: —BB*, wet ground below The Falls.

PONTEDERIACEÆ. PICKEREL-WEED FAMILY.

PONTEDERIA, L. PICKEREL-WEED.

P. cordata, L.
Shallow water. B, M, S and BB, common.

XYRIDACEÆ. YELLOW-EYED GRASS FAMILY.

XYRIS, Gronov. YELLOW-EYED GRASS.

X. flexuosa, Muhl.
Sandy shores, and boggy places. S*, edge of Turtle Pond.

Var. **pusilla,** Gray.
Sandy shores and boggy places. B, edge of Hoosicwhisick Pond.

X. Caroliniana, Walt.
Swamps and boggy pools. M, frequent.

JUNCACEÆ. RUSH FAMILY.

JUNCUS, Tourn. RUSH.

J. effusus, L. COMMON RUSH.
Wet ground. B and M, common: —S*, meadow near Office: —BB*, occasional.

J. marginatus, Rostk.
Moist, generally sandy ground. B*, occasional: —M*, meadow off Forest St.; damp wood-path, Melrose: —BB*, occurs.

J. Greenii, Oakes & Tuck.
Sandy soil near the coast. B, rare; summit of Nahanton Hill.

J. tenuis, Willd.
Dry fields and roadsides. B^*, common: — M^*, W. border of S. Reservoir, etc. : — S^*, near Bold Knob, etc. : — BB^*, occurs.

J. bufonius, L.
Damp ground. B^*, common.

J. pelocarpus, E. Meyer.
Wet sandy places. B, rare : — M^*, occasional; E. side of N. Reservoir, etc.

J. articulatus, L.
Wet ground. B^*, occurs : — M^*, meadow, W. of Forest St.; E. slope of Bear Hill.

J. acuminatus, Michx.
Low ground. B^* and BB^*, occurs : — M^*, swamp, S. of Black Rock, etc.

J. Canadensis, J. Gay, var. longicaudatus, Engelm.
Wet ground. B^*, occurs : — M^*, meadow, W. of Spot Pond, etc. A form between this and var. *subcaudatus*, Engelm, was collected at Hemlock Bound, B.

Var. coarctatus, Engelm.
Low ground. B^*, occasional : — M, E. side of N. Reservoir.

LUZULA, DC.   WOOD RUSH.

L. campestris, DC.
Dry woods, fields, clearings, and low ground. B, M^* and BB^*, common : — S^*, woods near Office.

TYPHACEÆ.   CAT-TAIL FAMILY.

TYPHA, Tourn.   CAT-TAIL FLAG.

T. latifolia, L. COMMON CAT-TAIL.
Swamps and bogs. B^*, frequent:—M, S and BB, common.

T. angustifolia, L.
Swamps and bogs. B^*, rare; near Sawcut Pass; near Monatiquot Stream.

SPARGANIUM, Tourn. Bur-reed.

S. simplex, Huds.
Shallow water, ditches and bogs. B^*, occasional; shore of Hoosicwhisick Pond; near Sawcut Notch, etc. : — M, frequent : — S, occasional.

Var. androcladum, Engelm.
Bogs and ditches, generally in shallow water. B^*, shore of Hoosicwhisick Pond.

ARACEÆ. Arum Family.

ARISÆMA, Mart. Indian Turnip.

A. triphyllum, Torr. Jack-in-the-Pulpit.
Rich shady woods, bogs, and wet meadows. B^* and M^*, common : — BB, occasional.

PELTANDRA, Raf. Arrow Arum.

P. undulata, Raf.
Wet meadows and bogs. B^*, frequent ; Cedar Swamp, etc. : — S^*, occurs : —BB^*, below The Falls.

CALLA, L. Water Arum.

C. palustris, L.
Bogs in shade. B^*, Cedar Swamp : — M, Virginia Wood.

SYMPLOCARPUS, Salisb. Skunk Cabbage.

S. fœtidus, Salisb.
Moist ground, bogs, and swamps. B, M^*, S and BB, common.

ACORUS, L. Sweet Flag.

A. Calamus, L.
Swamps and bogs. M, common in swamps : — BB^*, occurs.

LEMNACEÆ. Duckweed Family.

SPIRODELA, Schleid.

S. polyrrhiza, Schleid.

Ponds and pools, floating. *M*, frequent; border of Doleful Pond; shore of S. Reservoir; bog, N. side of Bear Hill.

LEMNA, L. Duckweed.

L. Valdiviana, Philippi.

Ponds and nooks in slow streams. *B**, this most interesting species grows in Monatiquot Stream, near Randolph Ave.

L. minor, L.

Ponds and pools, floating. *M*, common; bog, N. side of Bear Hill, etc.

ALISMACEÆ. Water-plantain Family.

ALISMA, L. Water-plantain.

A. Plantago, L.

Wet boggy places in soft mud. *B** and *S*, common: —*M*, ditch, N. E. of Bear Hill.

SAGITTARIA, L.† Arrowhead.

S. latifolia, Willd.‡   *S. variabilis*, Gray, Man. ed. 6, 554, in part.

Wet places. *B**, Cedar Swamp.

S. latifolia, Willd., form c.§   *S. variabilis*, Gray, Man. ed. 6, 554, in part.

Wet places. *B**, Balster Brook; Monatiquot Stream.

S. graminea, Michx.

Shallow water. *M**, common round Spot Pond.

---

†*S. variabilis*, Engelm. is frequent in ditches and wet meadows in *M*, but it has not been studied in the light of J. G. Smith's new revision.

‡ See 6th Ann. Rep. Missouri Bot. Garden, 34, t. 3.

§ See 6th Ann. Rep. Missouri Bot. Garden, 38, t. 5.

NAIADACEÆ. Pondweed Family.

POTAMOGETON, Tourn. Pondweed.

P. natans, L.
Ponds and slow streams. *M*, rare; Spot Pond.

P. Nuttallii, Ch. & Sch.† *P. Pennsylvanicus,* Gray, Man. ed. 6, 559.
Ponds and slow streams. *B*, Hoosicwhisick Pond: —*M*, abundant in ditches round Spot Pond.

P. Vaseyi, Robbins.
Ponds and pools. *M*, very rare; Spot Pond.

P. Spirillus, Tuck.
Ponds and slow streams. *B*, rare; Hoosicwhisick Pond: — *M**, W. shore of Spot Pond.

P. pusillus, L.
Ponds and slow streams. *B*, rare; ditch near Hoosicwhisick Pond.

NAIAS, L. Naiad.

N. flexilis, Rostk. & Schmidt.
Ponds and slow streams. *M**, border of N. Reservoir.

N. gracillima, Morong.‡ *N. Indica,* Cham., var. *gracillima,* Gray, Man. ed. 6, 566.

Ponds and slow streams. *M*, rare; Spot Pond.

ERIOCAULEÆ. Pipewort Family.

ERIOCAULON, L. Pipewort.

E. septangulare, With.
Borders of ponds and streams or in shallow water. *B**, Hoosicwhisick Pond, etc.: — *M*, occasional; Spot Pond and adjacent ditches: — *S**, Turtle Pond.

---

† See Mem. Torr. Bot. Club, iii. 18, t. 29.
‡ See Mem. Torr. Bot. Club, iii. 61, t. 68.

CYPERACEÆ. Sedge Family.

CYPERUS, Tourn. Galingale.

C. diandrus, Torr.
Low ground. B^*, border of Hoosicwhisick Pond : — M^*, by N. Reservoir.

C. filiculmis, Vahl.
Dry barren soil. B^* and BB^*, occurs.

C. dentatus, Torr.
Wet sandy places. B^*, border of Hoosicwhisick Pond : — M^*, shore of Spot Pond.

C. strigosus, L.
Low ground. B^* and BB^*, occurs : — M^*, shore of Spot Pond.

DULICHIUM, Pers.

D. spathaceum, Pers.
Pond borders and wet places. B^*, common : — M^*, occasional; border of Spot Pond, etc.

ELEOCHARIS, R. Br. Spike Rush.

E. tuberculosa, R. Br.
Sandy moist soil. B^*, rare; collected in two localities.

E. ovata, R. Br.
Muddy places, borders of ponds and ditches. B^*, border of Hoosicwhisick Pond : — M^*, common : — S^*, border of Turtle Pond, etc. : — BB^*, occurs.

E. Engelmanni, Steud.
Wet places. B^* and M, rare.

E. olivacea, Torr.
Wet sandy and muddy soil. B^*, rare : — M^*, meadow, W. of Spot Pond : — S^*, occasional; border of Turtle Pond, etc.

E. palustris, R. Br.
Wet places or in water. B, frequent; Hawk Hill, etc. : — M^*, meadow, Melrose.

E. tenuis, Schult.

Wet places, sometimes in rather dry sandy spots. $B*$, occurs : — M, frequent.

E. acicularis, R. Br.

Muddy places by ponds and streams; a delicate little plant in its flowering and fruiting form, but often filling ditches and streams with sterile culms, sometimes nearly a foot long, rarely accompanied by fertile culms; this peculiar form has long been unrecognizable. $B*$, border of Hoosicwhisick Pond: — $M*$, border of Spot Pond.

FIMBRISTYLIS, Vahl.

F. autumnalis, Roem. & Schult.

Low ground. $B*$, border of Hoosicwhisick Pond: — $M*$, rare; ditch in meadow, W. of Spot Pond.

F. capillaris, Gray.

Dry sandy soil. $B*$, frequent; Great Dome; top of Hancock Hill, etc.

SCIRPUS, Tourn. BULRUSH. CLUB RUSH.

S. planifolius, Muhl.

Dry or moist open woods. B, frequent on slope of Hawk Hill.

S. debilis, Pursh.

Wet swampy ground. $M*$, rare; W. shore of Spot Pond.

S. sylvaticus, L.

Wet ground, by brooks, etc. M, rare.

S. atrovirens, Muhl.

Wet meadows, bogs, and roadsides. $S*$, meadow, W. of Office.

ERIOPHORUM, L. COTTON GRASS.

E. cyperinum, L.

Wet roadsides, meadows, and swamps. $B*$ and $M*$, occurs.

E. vaginatum, L.

Cold bogs. $S*$, bog, W. of Turtle Pond.

E. Virginicum, L.

Wet meadows and low ground. *B**, frequent: —*M**, border of Doleful Pond: —*S**, border of Turtle Pond.

E. gracile, Koch.

Bogs. *M**, rare; meadow off Pond St.

RHYNCHOSPORA, Vahl. BEAK RUSH.

R. fusca, Roem. & Schult.

Low ground. *B*, in the S. E. part of the Reservation.

R. alba, Vahl.

Wet meadows and bogs. *B*, occurs: —*M**, frequent; meadow, N. E. of Bear Hill, etc.: —*S*, occasional.

R. glomerata, Vahl.

Low ground. *B**, shore of Hoosicwhisick Pond, etc.: — *M**, frequent; E. border of N. Reservoir, etc.

CAREX, Ruppius. SEDGE.

C. folliculata, L.

Swamps and wet meadows. *B**, common; Cedar Swamp; base of Hawk Hill, etc.

C. intumescens, Rudge.

Swamps and wet pastures. *B**, common: —*M*, occasional: —*S*, occurs.

C. lupulina, Muhl.

Wet places. *B**, common: —*M**, by Spot Pond, etc.

C. utriculata, Boott.

Swampy places. *M**, border of Doleful Pond.

Var. **minor,** Boott.

Swampy places. *M*, rare.

C. bullata, Schkuhr.

Low ground. *B*, occasional.

C. lurida, Wahl.

Wet places. *B**, common: —*M**, *S** and *BB**, occurs and doubtless common.

C. Pseudo-Cyperus, L., var. **Americana**, Hochst.
Swamps and wet places. *B*, frequent: —*M** and *S**, occurs.

C. scabrata, Schwein.
Wet meadows and bogs. *B**, rare; near Cragfoot Spring: —*M**, rare; boggy woods, Medford.

C. vestita, Willd.
Sandy places. *S**, meadow, W. of Office.

C. filiformis, L.
Boggy places. *M*, occasional.

Var. **latifolia**, Boeckl.
Bogs and low meadows. *BB**, occurs.

C. rigida, Gooden., var. **Goodenovii**, Bailey.† *C. vulgaris*, Gray, Man. ed. 6, 599.
Low meadows. *B*, occurs: —*M*, meadow near Spot Pond: —*S**, meadow, S. W. of Office.

C. rigida, Gooden., var. **strictiformis**, Bailey.‡ *C. vulgaris*, var. *strictiformis*, Gray, Man. ed. 6, 599.
Moist places. *B*, rare: —*M**, rare; border of swamp near Old Silver Mine.

C. stricta, Lam.
Low ground, meadows, and bogs. *M*, frequent: —*S** and *BB**, occurs.

Var. **angustata**, Bailey.
Low ground, meadows, and bogs. *M**, rare; meadow, N. E. of Bear Hill.

Var. **decora**, Bailey.
Low ground, meadows, and bogs. *BB*, by the brook.

C. crinita, Lam.
Meadows and damp roadsides. *B**, common: —*M**, shore of Spot Pond, etc.: —*S**, Happy Valley, etc.: —*BB**, occurs.

† See Gray, Man. ed. 6, 2d issue, 735 c.
‡ See Jour. Bot. xxviii. 172 (1890).

C. virescens, Muhl.

Rich shady ground. *B**, common: — *M**, pine woods off Ravine Road, etc.; slender form near Cascade: — *S**, swamp, S. W. of Office.

Var. **costata**, Dewey.

Rich shady ground. *B*, frequent; S. slope of Great Blue Hill, etc.

C. debilis, Michx., var. **Rudgei**, Bailey.

Moist ground. *B*, occasional: — *M**, Pine Hill; — *S**, swamp, S. W. of Office: — *B B**, occurs.

C. gracillima, Schwein.

Meadows and open woods. *B*, *S** and *B B**, occurs: — *M**, Virginia Wood; Pine Hill, etc.

C. flava, L., var. **graminis**, Bailey.

Wet meadows and grassy places. *M**, rare; E. shore of Spot Pond.

C. pallescens, L.

Meadows and damp places. *B*, frequent; Hawk Hill, etc.: — *M**, *S* and *B B**, occurs.

C. conoidea, Schkuhr.

Moist places. *M**, rare; meadow, N. E. of Bear Hill; meadow, N. end of Doleful Pond.

C. laxiflora, Lam., var. **blanda**, Boott.† *C. laxiflora*, var. *striatula*, Gray, Man. ed. 6, 607.

Open woods and grassy places. *M**, Pine Hill, etc.: — *B B**, occurs.

Var. **latifolia**, Boott.

Rich woods. *B*, rare; Rattle Rock.

Var. **patulifolia**, Carey.

Open woods and moist places. *B*, frequent; Hawk Hill, etc.: — *M**, Virginia Wood; Pine Hill; Cascade region, etc.: — *S*, occurs.

† See Coulter, Contrib. U. S. Nat. Herb. ii. 481.

C. digitalis, Willd.

Open woods. *B*, occasional; Great Blue Hill, etc. : — *M**, wood-road, W. of Pine Hill.

C. laxiculmis, Schwein.

Open woods. *B* and *B B**, rare :—*M**, rare; damp woods off Ravine Road.

C. platyphylla, Carey.

Rich woods. *M*, rare.

C. panicea, L.

Nat. from Eu. in fields and grassy places. *B*, occasional; Hawk Hill, etc.

C. varia, Muhl.

Dry woods and open places. *M**, Virginia Wood; border of thicket, N. side of Spot Pond : —*S**, upland rocks.

C. Novæ-Angliæ, Schwein.

Swamps, low ground, and shady banks. *B*, "Blue Hills, June 3, 1870," coll. William Boott and in Gray Herb.†

C. Pennsylvanica, Lam.

Dry slopes and fields; one of our earliest and commonest species. *B* and *B B**, common: —*M** and *S*, occurs and doubtless common.

C. communis, Bailey.

Dry open woods and hillsides. *B*, rare; near Nahanton Hill.

C. præcox, Jacq.

Nat. from Eu. in fields. *S*, rare; Bellevue Hill.

C. umbellata, Schkuhr.

Dry slopes and woods. *B**, frequent; Wampatuck Hill; slope of Great Blue Hill, etc. : —*M** and *S**, common.

C. polytrichoides, Muhl.

Moist open woods and bogs. *B* and *M*, frequent.

---

† Collected also by Wm. Boott, July 10, 1861, in Purgatory Swamp, Dedham, near the Reservation. The specimen is in the Gray Herb.

C. stipata, Muhl.

Moist open ground. *B*, common: —*M**, frequent: —*S** and *BB**, occurs and doubtless common.

C. vulpinoidea, Michx.

Low ground. *M**, damp open places in woods, Stoneham: —*S**, swamp, N. W. of Office.

C. tenella, Schkuhr.

Swamps and cold woods. *B*, occasional in Cedar Swamp.

C. rosea, Schkuhr.

Rich woods. *B*, frequent at base of Great Blue Hill: —*M**, Pine Hill: —*S*, occurs.

Var. **radiata,** Dewey.

Open dry woods and sunny places. *B*, occasional: —*M*, Cascade.

Var. **retroflexa,** Torr.

Open places and dry woods. *M**, open rocky woods, Stoneham: —*S**, upland woods.

C. sparganioides, Muhl.

Rich woods. *B* and *M*, occasional.

C. muricata, L.

Nat. from Eu. in dry fields and on sunny slopes. *M**, low ground off Ravine Road, "The nerves unusually strong," L. H. Bailey: —*BB**, occurs.

C. Muhlenbergii, Schkuhr.

Open sterile ground. *B*, rare; Rattlesnake Hill.

C. cephalophora, Muhl.

Dry soil. *B* and *BB**, occurs: —*M**, W. border of S. Reservoir; wood-road, N. of Pine Hill.

C. sterilis, Willd.† *C. echinata,* var. *microstachys,* Gray, Man. ed. 6, 618.

Wet meadows, and low ground. *M**, occasional; N. edge of Doleful Pond, etc.: —*S**, by Turtle Pond: —*BB**, occurs.

† See Bull. Torr. Bot. Club, xx. 424.

Var. **excelsior**, Bailey. †
Wet bogs and damp woods. M^*, border of Doleful Pond; bog, N. side of Bear Hill; a form approaching var. *cephalantha*, Bailey, wet meadow, N. of Bear Hill:—S^*, border of Turtle Pond, etc.

Var. **cephalantha**, Bailey.‡ *C. echinata*, var. *cephalantha*,
Gray, Man. ed. 6, 618.
Meadows and low ground. M, ditch, N. of Spot Pond.

C. **Atlantica**, Bailey. § *C. echinata*, var. *conferta*, Gray,
Man. ed. 6, 618.
Bogs and low ground. S^*, rare; S. slope of Bellevue Hill; damp roadside.

C. **interior**, Bailey, var. **capillacea**, Bailey. ‖
Dryish bogs. B^*, rare: — S, border of Turtle Pond.

C. **canescens**, L.
Low ground, bogs, and swamps. B and M^*, common: — S^*, meadow near Office:— BB^*, occurs.

Var. **vulgaris**, Bailey.
Generally in drier places. M^*, border of bog, N. side of Bear Hill.

C. **trisperma**, Dewey.
Damp woods and cold bogs. B^*, occasional; Cedar Swamp, etc.:—S^*, swamp, W. of Turtle Pond.

C. **bromoides**, Schkuhr.
Wet open woods. M^*, Pine Hill; boggy woods off Forest St.:— BB^*, boggy woods.

C. **tribuloides**, Wahl.
Open low ground. B and BB^*, occurs.

Var. **reducta**, Bailey.
Open low ground. BB, occurs.

---

† See Bull. Torr. Bot. Club, xx. 424.
‡ See Bull. Torr. Bot. Club, xx. 425.
§ See Bull. Torr. Bot. Club, xx. 425.
‖ See Bull. Torr. Bot. Club, xx. 426.

C. scoparia, Schkuhr.

Open places in meadows and by roadsides. *B*, M** and *S*,* occurs. Doubtless common in all the Reservations.

C. fœnea, Willd.

Dryish open places. *B**, occasional : — *M**, occasional; border of S. Reservoir; N. end of Doleful Pond, etc. : —*BB**, in shade near Trapelo Road.

C. straminea, Willd.

Dryish fields and open woods. *M**, Pine Hill.

Var. **mirabilis,** Tuck.

Shady places. *B*, frequent at base of Hawk Hill : —*S**, damp roadside and meadows : — *B B*, occurs.

Var. **aperta,** Boott.

Bogs and wet places. *B*, rare.

GRAMINEÆ. Grass Family.

PASPALUM, L.

P. setaceum, Michx.

Sandy and gravelly places. *B**, near Hoosicwhisick Pond; valley, E. of Wampatuck Hill, etc. :—*M**, occasional :— *BB**, occurs.

PANICUM, L.† Panic Grass.

P. sanguinale, L. Crab Grass.

Nat. from Eu. in cult. ground and waste places. *B**, slope of Great Blue Hill, etc.

P. glabrum, Gaudin.

Nat. from Eu. in cult. ground and waste places. *B**, occasional.

P. filiforme, L.

Dry sandy soil. *B**, rare; gravelly road, W. Quincy.

† The arrangement of the species of this genus is acc. to Scribner (Grasses of Tenn. pt. 2), who kindly examined the difficult *dichotomum* group.

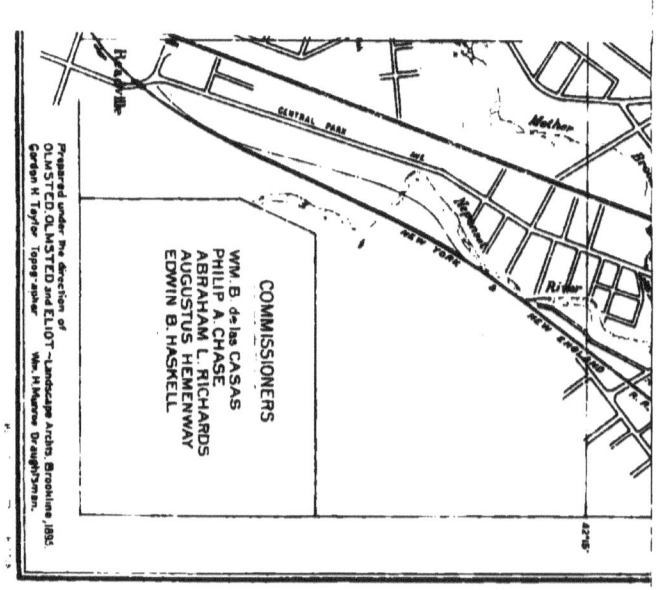

P. Crus-galli, L. Barnyard Grass.
Nat. from Eu. in moist waste places. *B**, occasional: — *M*, frequent.

P. miliaceum, L.† True Millet.
Cult. and rarely escaped. *B*, rare.

P. agrostoides, Muhl.
Wet meadows and borders of streams. *B**, occasional; border of Hoosicwhisick Pond, etc.: — *M**, occasional; E. border of N. Reservoir, etc.

P. proliferum, Lam.
Borders of ponds and streams. *B**, rare; by pool in Pine Tree Brook.

P. capillare, L.
Cult. fields and all waste places. *B B**, occurs; doubtless common in all the Reservations.

P. latifolium, L.
Low ground in shady places. *B**, common: — *M**, occurs: — *S**, Rooney's Rock, etc.

P. commutatum, Schult.
Rocky woods and hillsides. *M**, N. of Cascade.

P. scoparium, Lam.
Wet meadows and damp places. *B**, occurs.

P. depauperatum, Muhl.
Dry woods and sandy places. *B**, occurs: — *M**, near Cascade, etc.: — *S**, Milkweed Hill, etc.

P. sphærocarpon, Ell. *P. nitidum*, Gray, Man. ed. 6, 632, in part.
Dry or moist ground. *M**, wet borders of N. and W. sides of S. Reservoir.

P. lanuginosum, Ell. *P. dichotomum*, Gray, Man. ed. 6, 633, in part.
Open woods and thickets, usually in moist soil; generally

† See Vasey, Contrib. U. S. Nat. Herb. iii. 34.

cæspitose; hairy throughout; panicle two or three inches long, diffuse; spikelets three-fourths to nearly one line long, oblong-elliptical, obtuse; second and third glumes seven-nerved, pubescent; fourth glume smooth, obtuse; common in this region. *M**, borders of Reservoirs and Doleful Pond; rocky woods, meadows, etc.: — *S**, Happy Valley; Bellevue Hill, etc.: — *B B**, occurs.

P. nitidum, Lam.

Open places. *B*, occasional: — *M**, borders of Spot Pond, Doleful Pond, N. Reservoir, etc.

P. dichotomum, L.

Dry or wet soil, in sun or shade. *B**, common: —*M**, border of Doleful Pond; hillside, N. of Cascade, etc.: — *S**, Milkweed Hill; Happy Valley.

P. barbulatum, Michx. *P. dichotomum,* Gray, Man. ed. 6, 633, in part.

Open woods; spikelets smooth; nodes bearded with reflexed hairs. *B**, occurs.

SETARIA, Beauv. BRISTLY FOXTAIL GRASS.

S. GLAUCA, Beauv. FOXTAIL GRASS.

Adv. from Eu. in fields and waste places. *B** and *M*, occurs.

S. VIRIDIS, Beauv. GREEN FOXTAIL.

Adv. from Eu. in fields and waste places. *B**, occasional: — *M**, occurs.

LEERSIA, Swartz. WHITE GRASS.

L. Virginica, Willd. WHITE GRASS.

Wet shady places. *B**, occasional: — *S* and *B B**, occurs.

L. oryzoides, Swartz. RICE CUT GRASS.

Wet places. *B**, border of Hoosicwhisick Pond, etc.: — *B B**, occurs.

MISCANTHUS, Anders.

M. SINENSIS, Anders. *Eulalia Japonica,* forma *variegata,* of nurserymen. ZEBRA GRASS.

CATALOGUE OF PLANTS. 99

Cult. from Japan. *B*, old garden, Park's place, Hawk Hill.

ANDROPOGON, Royen. BEARD GRASS.

A. furcatus, Muhl.
Dry barren soil. *B**, Great Blue Hill; E. slope of Hancock Hill.

A. scoparius, Michx.
Dry fields and roadsides. *B**, common: — *M*, occurs and doubtless common: — *S*, frequent.

A. Virginicus, L.
Sandy places. *B*, rare.

CHRYSOPOGON, Trin.

C. nutans, Benth.
Dry ground. *B**, Hillside St.; near the top of Hancock Hill.

Under ANTHOXANTHUM insert — A. ODORATUM, L.

HIEROCHLOE, Gmelin. SWEET GRASS.

H. borealis, Roem. & Schult.
Wet chiefly brackish meadows, and salt marshes. *M**, rare; N. side of Spot Pond.

ARISTIDA, L. TRIPLE-AWNED GRASS.

A. dichotoma, Michx. POVERTY GRASS.
Dry sandy or gravelly places. *B**, occasional; dry paths on Great Blue Hill, etc.

A. gracilis, Ell.
Sandy places. *B**, valley, E. of Wampatuck Hill, etc.

STIPA, L. FEATHER GRASS.

S. avenacea, L. BLACK OAT GRASS.
Dry woods. *B*, rare; base of Great Blue Hill.

ORYZOPSIS, Michx. MOUNTAIN RICE.

O. melanocarpa, Muhl.
Dry rocky woods. *M**, rare; Cascade.

O. asperifolia, Michx.
Rich woods. *B*, rare; N. base of Hawk Hill: — *M**, rare.

O. Canadensis, Torr.
Rocky slopes and dry places. *B*, rare; Wampatuck Hill: — *S*, common.

MUHLENBERGIA, Schreb. DROP-SEED GRASS.

M. sobolifera, Trin.
Rocky places in open woods. *B*, occasional; base of Hancock Hill.

M. glomerata, Trin.
Boggy places and damp roadsides. *B**, rare; border of pool, E. of Great Dome: — *S*, rare.

M. Mexicana, Trin.
Moist places. *B*, occurs.

M. sylvatica, Torr. & Gray.
Rocky or low woods and shady places. *B*, frequent; S. slope of Great Blue Hill, etc.: — *M*, occasional.

M. Willdenovii, Trin.
Rocky shady places. *B**, bases of Hawk Hill, Great Blue Hill, etc.: — *M**, occasional; W. border of Spot Pond, etc.

BRACHYELYTRUM, Beauv.

B. aristatum, Beauv.
Rocky woods. *B**, occurs: — *S**, Happy Valley.

PHLEUM, L. CAT'S-TAIL GRASS.

P. PRATENSE, L. TIMOTHY. HERD'S GRASS.
Nat. from Eu. in meadows and by roadsides. *B* and *S**, occurs: — *M**, common. Doubtless common in all the Reservations.

ALOPECURUS, L. Fox-tail Grass.

A. PRATENSIS, L.   MEADOW FOX-TAIL.
Nat. from Eu. in meadows, pastures, and by roadsides.
M, frequent: — *BB**, occurs.

A. GENICULATUS, L.   FLOATING FOX-TAIL.
Nat. from Eu. in meadows and moist places. *M**, swamp, slope of Bear Hill.

Var. **aristulatus**, Torr.
Wet places. *M**, swamps on slopes of Bear Hill; E. shore of Spot Pond.

SPOROBOLUS, R. Br.   DROP-SEED GRASS.

S. **vaginæflorus**, Vasey.
Dry and barren places. *B**, occurs.

S. **serotinus**, Gray.
Wet places. *B**, common: — *M**, meadow, base of Bear Hill.

AGROSTIS, L.   BENT GRASS.

A. ALBA, L.   WHITE BENT GRASS.
Nat. from Eu. and cult. *M**, occurs.

Var. VULGARIS, Thurb.   RED TOP.
Nat. from Eu. and cult. *M** and *BB**, occurs: — *M**, forma *aristata* occurs.

A. **perennans**, Tuck.   THIN GRASS.
Moist shaded places. *B**, *M** and *S* *, occurs and doubtless common.

A. **scabra**, Willd.   HAIR GRASS.
Dry places. *B**, occurs: — *M**, frequent.

A. **canina**, L.   BROWN BENT GRASS.
Meadows and moist places. *S**, meadow, W. of Office.

CINNA, L.   WOOD REED GRASS.

C. **arundinacea**, L.

Moist shady woods and swamps. *B**, *M** and *S*, occasional.

CALAMAGROSTIS, Adans. REED BENT GRASS.

C. Canadensis, Beauv. BLUE JOINT.

Wet places. *M*, frequent: — *S**, occurs and doubtless common.

C. Nuttalliana, Steud.

Moist places. *B**, near Cedar Swamp: — *M**, border of wood-path, Melrose: — *S*, rare.

HOLCUS, L. MEADOW SOFT GRASS.

H. LANATUS, L. VELVET GRASS.

Nat. from Eu. in damp meadows. *B*, found in a few localities near Hawk Hill.

DESCHAMPSIA, Beauv.

D. flexuosa, Trin.

Dry fields and hillsides. *B*, common.

AVENA, Tourn. OAT.

A. SATIVA, L. COMMON OAT.

Cult. from Old World, and escaped. *B**, frequent along roadsides.

DANTHONIA, DC. WILD OAT GRASS.

D. spicata, Beauv.

Dry fields and sterile soil. *B**, common: — *M** and *S**, occurs. Doubtless common in all the Reservations.

D. compressa, Aust.

Dry open places. *B**, occasional: — *M**, near Bear's Den path and Cascade, etc.

EATONIA, Raf.

E. Dudleyi, Vasey.

Meadows and moist land. *M*, Cascade region.

ERAGROSTIS, Beauv.

E. capillaris, Nees.
Dry fields and open woods. *B*, rare; ledge on S. slope of Great Blue Hill.

E. pectinacea, Gray.
Dry fields and sandy places. *B*, occasional.

DACTYLIS, L. ORCHARD GRASS.

D. GLOMERATA, L.
Nat. from Eu. in fields and by roadsides. *B* and *M*, common: — *B B**, occurs.

POA, L. SPEAR GRASS. MEADOW GRASS.

P. ANNUA, L. LOW SPEAR GRASS.
Nat. from Eu. in cult. and waste grounds. *B*, frequent:— *M*, common in fields and by roadsides: — *B B**, occurs.

P. COMPRESSA, L. WIRE GRASS.
Nat. from Eu. in fields and waste places. *B**, frequent:— *M*, common: — *S ** and *B B**, occurs and doubtless common.

P. serotina, Ehrh. FALSE RED-TOP. FOWL MEADOW-GRASS.
Low ground. *M**, common: — *S **, occurs.

P. pratensis, L. JUNE GRASS. KENTUCKY BLUE GRASS.
Meadows, pastures, and roadsides. *B**, common: — *M**, *S ** and *B B**, occurs and doubtless common.

GLYCERIA, R. Br. MANNA GRASS.

G. Canadensis, Trin. RATTLESNAKE GRASS.
Wet places. *B** and *M*, frequent: — *S*, occasional.

G. obtusa, Trin.
Boggy places. *B**, frequent; Cedar Swamp; near Hoosicwhisick Pond, etc.: — *S **, occasional; border of Turtle Pond, etc.

G. nervata, Trin.
Meadows and wet open places. *B** and *M**, frequent: — *S ** and *B B**, occurs.

G. pallida, Trin.
Shallow water. *M**, frequent; border of Wright's Pond; near Spot Pond, etc.

G. fluitans, R. Br.
Shallow water. *B**, occurs: — *M**, rare; S. side of Spot Pond: — *S**, swamp, S. W. of Office.

G. acutiflora, Torr.
Shallow water and wet places. *M**, occasional; brook, W. of Bear Hill; swamp, S. of Black Rock, etc.

FESTUCA, L. FESCUE GRASS.

F. tenella, Willd.
Dry barren places. *B**, occasional; Hawk Hill, etc.: — *S**, rare; Bold Knob.

F. ovina, L. SHEEP'S FESCUE.
Pastures. *BB**, rare.

F. nutans, Willd.
Rocky places in woods. *B*, frequent; Hawk Hill, etc.: — *M**, rare; foot of Cascade.

F. ELATIOR, L., var. PRATENSIS, Gray.
Nat. from Eu. in grassy places. *M*, frequent: — *S** and *BB**, occurs. Doubtless common in all the Reservations.

BROMUS, L. BROME GRASS.

B. SECALINUS, L. CHESS. CHEAT.
Adv. from Eu. in wheat fields and waste places. *B**, house-sites: — *M**, rare; single plant in pasture.

B. RACEMOSUS, L. UPRIGHT CHESS.
Adv. from Eu. in waste places. *B*, rare; Chickatawbut Hill.

B. ciliatus, L.
Moist or dry shady places. *B*, frequent: — *M**, Cascade; rocky woods, slope of Bear Hill, etc.: — *S**, foot of Overbrook Hill.

AGROPYRUM, Gaertn. FALSE WHEAT.

A. repens, Beauv. WITCH GRASS.
Cult. and waste places everywhere. *M**, frequent. Doubtless in all the Reservations.

A. caninum, Roem. & Schult.
Rocky woods and cult. ground. *M**, rare; rocky woods, Stoneham.

TRITICUM, L. WHEAT.

T. SATIVUM, Lam. COMMON WHEAT.
Cult. and escaped. *B**, roadside by Wolcott Pines, the short-awned form.

ELYMUS, L. WILD RYE. LYME GRASS.

E. Canadensis, L.
Banks of streams and wet places. *B*, occasional.

E. striatus, Willd.
Rocky woods and shade. *B*, occasional.

ASPRELLA, Willd. BOTTLE BRUSH GRASS.

A. Hystrix, Willd.
Rich woods. *B*, occasional: — *M*, rocks below Cascade.

CLASS II. GYMNOSPERMÆ.

CONIFERÆ. PINE FAMILY.

PINUS, Tourn. PINE.

P. Strobus, L. WHITE PINE.
Moist ground and hillsides. *B*, common throughout in the valleys and on the hill-tops; the largest white pines in the Reservation are among the Crossman Pines; the finest grove is near the junction of Canton and Blue Hill Avenues: — *M**, common; large trees at the S. end of Spot Pond and W. edge of the Reservation: — *S*, common; but few large trees in the

Reservation: — *B B*, occasional; a few large trees in the N. section.

P. rigida, Mill. PITCH PINE.

Sterile or sandy soil. *B*, frequent throughout; good trees in an old pasture, W. of Balster Brook; a few on the top of Great Blue Hill and Rattlesnake Hill, etc. : — *M*, common : — *S*, frequent : — *B B*, frequent; a group on the ridge, N. of the Waverly Oaks; several old trees near Trapelo Road, in the N. section.

P. SYLVESTRIS, L. SCOTCH PINE.

Cult. from Eu. *B B*, an old tree near superintendent's house.

PICEA, Link. SPRUCE.

P. EXCELSA, Link. NORWAY SPRUCE.

Planted from Eu.; the commonest and most vigorous species. *B* and *M*, occasional about house-sites : — *B B*, by superintendent's house.

TSUGA, Carr. HEMLOCK.

T. Canadensis, Carr.

Dry rocky woods and ridges, and in rich soil by borders of streams. *B*, frequent; outlet of Blueberry Swamp; Bear Hill; Hancock Hill, etc.; a fine specimen is perched upon the top of a massive square bowlder, 6 ft. or more above the ground, and forms the town boundary between Quincy and Braintree, its diameter being 5 ft. 10½ in.; " The E. slope of South-east Ridge was a hemlock forest, about 30 years ago. The trees were blazed to be cut. Peleg Bronson, then a boy, while passing through the woods, marked some of the best of them, ' Do not cut, to be reserved,' not with authority, but hoping thus to save them. The trees are still living, decrepit old patriarchs " : — *M*, frequent throughout; large and tall trees at the S. end of Spot Pond; a grove of large trees near Ravine Road.

CHAMÆCYPARIS, Spach. WHITE CEDAR. CYPRESS.

C. sphæroidea, Spach.

Cold deep swamps. *B*, frequent in several swamps in W.

Quincy, notably in Cedar Swamp, where some large trees still survive : — *M*, rare ; confined to Cedar island in Cedar Swamp, and the lowland, W. of Bear Hill.

JUNIPERUS, L. JUNIPER.

J. communis, L. COMMON JUNIPER.

Dry sterile ground. *B*, frequent; more confined to pastures than elsewhere, but shaded out by growing woods; pasture, E. of Balster Brook ; base of Chickatawbut Hill, etc. : — *M**, frequent in rocky upland woods and fields : —*S* and *B B*, occasional in pastures.

J. Virginiana, L. RED CEDAR. SAVIN.

Dry ridges, hills, and pastures. *B**, common; grows naturally on the sunny hill-tops and in the open pastures, but is easily shaded out by growing trees ; fine trees, 20 or 30 ft. high in the pasture, E. of Pine Tree Brook ; summit on the W. side of Great Blue Hill ; a group on the summit of Chickatawbut Hill, etc. : — *M**, common on rocky wooded hillsides ; abundant on the summit and upper slopes of Bear Hill : —*S* and *BB*, common.

SERIES II. CRYPTOGAMIA. FLOWERLESS PLANTS.

CLASS I. PTERIDOPHYTA.

(By George E. Davenport.)

The diverse character of the Reservations justifies expectations of finding a liberal representation of the vascular cryptogams in their flora, and these expectations will no doubt be fully realized when all parts of the Reservations shall have been thoroughly worked over. As the present report is based wholly upon specimens actually collected, or known to exist in reliable herbaria, a necessarily somewhat incomplete report must be expected. An examination of this report will show that the Blue Hills and Middlesex Fells Reservations have been more thoroughly worked over than the Stony Brook and Beaver Brook Reservations, although the latter can scarcely be any the less interesting to a botanist studying the Ferns and their Allies.

The arrangement adopted here is that of the late Prof. D. C. Eaton, in the 6th edition of Gray's Manual, and comments are made in connection with the different species for the purpose of calling attention to certain points that may lead to a more thorough investigation, and result in restoring under proper guidance many species once common, or fairly plentiful, but now nearly or wholly extirpated from various causes now happily within control and future prevention.

EQUISETACEÆ. Horsetail Family.

EQUISETUM, L. Horsetail. Scouring Rush.

E. arvense, L. Common Horsetail.

Common everywhere in sandy soil; the fertile stems appearing very early in spring and soon perishing, the sterile stems appearing later and lasting all summer. B and M, common:— BB^*, near the ponds.

E. sylvaticum, L.

A rather scarce plant growing in wet shady woodlands; the fertile stems fruiting early in spring, the fruit soon withering away, but the stems remaining all summer with the sterile. B^*, occurs.

E. limosum, L.

A rare plant; stems all of one kind, in shallow water, fruiting in summer. M^*, occasional; ditch, S. of Bear Hill; meadows, N. side of Spot Pond, etc.: — S^*, meadow near Office.

E. hyemale, L. Scouring Rush.

Wet places, evergreen. M^*, occurs: — BB, meadow near the brook.

FILICES. Ferns.

POLYPODIUM, L. Polypody.

P. vulgare, L. Common Polypody.

Rocks. *B**, common: — *M**, abundant wherever there are ledges; seen growing on a large tree near Cascade some years ago: — *S*, reported as common on damp rocks in woods.

ADIANTUM, L. Maidenhair.

A. pedatum, L.

Rocky ravines. *B**, occasional; Hawk Hill; Hancock Hill, etc.: — *M*, occasional within the past twenty years, but now exceedingly rare; a fern so beautiful as this should not be permitted to become extinct when it can be so readily replenished from regions where it is abundant, and then carefully protected.

PTERIS, L. Brake.

P. aquilina, L.

Thickets and dry knolls. *B**, *M*, *S* and *BB*, common.

WOODWARDIA, Smith. Chain Fern.

W. Virginica, Smith.

Swamps and woodlands. *B**, rare; near Monatiquot Stream, etc.: — *M**, frequent in the S. part of the Reservation.

W. angustifolia, Smith.

Low woodlands; rare. *B**, rare; near Monatiquot Stream, W. Quincy; Cedar Swamp.

ASPLENIUM, L. Spleenwort.

A. Trichomanes, L. False Maidenhair.

Crevices of rocky declivities; becoming rare. *B*, frequent on ledges near top of Great Blue Hill: — *M**, occasional; rocks, N. side of Spot Pond; Cascade region; rocky woods, Bear Hill.

A. ebeneum, Ait. EBONY SPLEENWORT.

Occasional on knolls and ledges. *B**, frequent; by Monatiquot Stream; rocky places on Great Blue Hill, etc. : — *M**, occasional; Cascade region; abundant on Bear Hill, etc.

A. thelypteroides, Michx.

Rich woods. *B*, rare; near Cedar Swamp : — *M*, found in the Pine Hill woodlands, some years ago.

A. Filix-fœmina, Bernh. LADY FERN.

Common in a variety of forms in moist woodlands, the var. *angustum*, D. C. Eaton, in dry exposed situations. *B**, frequent; by Old Furnace Brook; N. and E. of Babel Rock, etc.; var. *angustum*, D. C. Eaton, slope of Hawk Hill : — *M**, common in damp woods : — *B B*, occurs.

PHEGOPTERIS, Fée. BEECH FERN.

P. polypodioides, Fée.

Occasional in damp woods and ravines. *B**, occurs : — *M*, rare.

P. hexagonoptera, Fée.

Damp woods; rare. *B**, rare; base of Hawk Hill, etc. : — *M**, rare; damp woods, E. slope of Bear Hill, and two stations in Medford.

P. Dryopteris, Fée.

Damp rocky woods; rare. *M*, reported as abundant in one locality, Melrose.

ASPIDIUM, Swz. SHIELD FERN.

A. Thelypteris, Swz.

Common in various situations in low ground. *B** and *M**, common.

A. simulatum, Davenport.†

Low moist woodlands with *A. Thelypteris* and *A. Noveboracence*; it thrives best in shaded situations, fruiting abundantly where *A. Thelypteris* is nearly always sterile; as yet not well

† See Bot. Gaz. xix. 495.

known to our limits, but turning up frequently elsewhere. *B**, rare; Cedar Swamp:—*M*, two fronds from Stoneham, probably in the Reservation limits near Bear Hill, in the Hitching collection in the Appalachian Club Herb.

A. Noveboracense, Swz. WOOD FERN.

Common in various situations in low ground. *B** and *M**, common. Doubtless in *S* and *BB*.

A. spinulosum, Swz.

Damp woods. *B** and *M*, occurs.

Var. **intermedium**, D. C. Eaton.

Swamps, with the type. *M**, occasional; Pine Hill woods, etc.

Var. **dilatatum**, Hook.

With the type. *B** and *M*, occurs.

A. Boottii, Tuck.

Swamps with *A. spinulosum* and *A. cristatum*. *B**, rare; damp woods:—*M**, rare; Pine Hill swamp, etc. One of our choicest ferns, never plentiful anywhere, and should be carefully protected.

A. cristatum, Swz.

Swamps; not plentiful anywhere. *B**, damp woods, W. Quincy, etc.:—*M**, Pine Hill swamp.

Var. **Clintonianum**, D. C. Eaton.

Low swamps with the type; rare. *M**, Medford.

A. cristatum × marginale, Davenport.†

Rocky and swampy woodlands; should be looked for wherever the parent species, *cristatum* and *marginale*, are found. *M**, Pine Hill swamp; two plants only were found, and the spot has since been probably destroyed.

A. marginale, Swz.

Rocky woods and swamps; frequent. *B**, frequent; Hawk Hill, etc.:—*M** and *S*, common in woods.

† See Bot. Gaz. xix. 494.

A. acrostichoides, Swz.

Rocky woodlands. B^*, frequent: — M^*, common: — S, occasional. A form near, but not quite deeply enough cut for a type of var. *incisum*, Gray, was collected in B.

CYSTOPTERIS, Bernh. BLADDER FERN.

C. fragilis, Bernh.

Moist rocks and shaded cliffs; occasional. M^*, frequent; crevices of moss-covered rocks, Virginia Wood; rocky border of Spot Pond, etc.

ONOCLEA, L.†

O. sensibilis, L. SENSITIVE FERN.

Common everywhere in low meadows and thickets. B^* and M, reported as common in meadows and damp woods. Doubtless common in S and BB.

WOODSIA, R. BR.

W. Ilvensis, R. Br.

Occasional on exposed ledges. B^*, rare :—M, frequent on rocks in upland woods.

W. obtusa, Torr.

Rocky woods, crevices, and pockets of ledges; rare. M^*, occasional; W. side of S. Reservoir; side of Bear Hill, etc.

DICKSONIA, L'Her.

D. pilosiuscula, Willd.

Common in shady woodlands and low grounds. B^*, common:—M^* and S, frequent.

OSMUNDA, L. FLOWERING FERN.

O. regalis, L. FLOWERING FERN.

Common in low woodlands. B^* and M, common:—S, damp woods.

---

† The noble Ostrich Fern, *O. Struthiopteris*, Hoffm., formerly grew very near the present limits of the Fells, and should be introduced into all the Reservations.

VITACEÆ. Vine Family.

VITIS, Tourn. Grape.

V. Labrusca, L. Northern Fox Grape.

Moist and rocky woods. *B*, occasional in moist places; slope of Chickatawbut Hill, etc.; common by Monatiquot Stream: — *M*, common in rocky woods and low thickets.

V. æstivalis, Michx. Summer Grape.

Thickets. *B*, frequent: — *M**, occasional in rocky and low woods: — *S**, E. of Rooney's Rock.

AMPELOPSIS, Michx. Virginian Creeper. Woodbine.

A. quinquefolia, Michx.

Rocky moist woods and sunny ledges. *B**, common in moist runs, on edges of swamps, and in rocky woods: — *M*, common on rocks and trees in woods: — *S* and *B B*, occasional.

SAPINDACEÆ. Soapberry Family.

ÆSCULUS, L. Horse-Chestnut. Buckeye.

Æ. Hippocastanum, L. Common Horse-Chestnut.

Cult. from Asia as a shade tree, and occasionally escaped. *B**, house-site, Hillside St.; Great Blue Hill, single tree.

ACER, Tourn. Maple.

A. Pennsylvanicum, L. Striped Maple.

Rich woods. *B*, rare.

A. saccharinum, Wang. Sugar or Rock Maple.

Rich woods. *B**, occasional on slopes near streams and in rocky glens; Witch Hazel Run; E. of Hawk Hill, etc.; rarely a good sized tree; planted along Hillside St.:—*M**, common as a shrub or small tree in rocky woods:—*S*, frequent:—*B B*, cult. near the superintendent's house.

A. dasycarpum, Ehrh. White or Silver Maple.

River banks and wet places. *B*, planted on Randolph Ave.

A. rubrum, L. RED OR SWAMP MAPLE.

Swamps and wet woods. *B*, common in wet ground and occasional on hill slopes; the finest single trees are on the knoll on S. side of Pine Tree Brook valley:—*M**, damp woods throughout; fine trees just W. of Pine Hill:—*B B*, occasional.

A. PLATANOIDES, L. NORWAY MAPLE.

Cult. from Eu. as a shade tree. *B*, planted by house, Canton Ave.

ANACARDIACEÆ. CASHEW FAMILY.

RHUS, L. SUMACH.

R. typhina, L. STAGHORN SUMACH.

Hillsides and thickets. *B*, common throughout:—*M*, common in upland rocky woods.

R. glabra, L. SMOOTH SUMACH.

Open woods and rocky soil. *B**, common throughout:—*M**, common in open woody places:—*S*, frequent; Milkweed Hill, etc.:—*B B**, occurs.

R. copallina, L. DWARF SUMACH.

Rocky hills, dry fields, and ledges. *B**, common throughout:—*M*, occasional in old pastures and open ground:—*S*, occasional.

R. venenata, DC. POISON SUMACH. DOGWOOD.

Swamps. *B**, frequent; Cedar Swamp; meadows by Monatiquot Stream; Blueberry Swamp, etc.:—*M*, frequent in boggy woods:—*S*, border of Turtle Pond.

R. Toxicodendron, L. POISON IVY. POISON OAK.

Fields, roadsides, and dry or wet woods; soon killed by fire. *B** and *S*, common in all kinds of soil, in shade or sun: —*M*, common in thickets and low grounds:—*B B*, common.

POLYGALACEÆ. MILKWORT FAMILY.

POLYGALA, Tourn. MILKWORT.

P. polygama, Walt.

Dry sandy soil. *B**, occasional; Hancock Hill; near Hoosicwhisick Pond, etc.: — *M**, common in open places in dry woods.

P. sanguinea, L.

Sandy and moist ground. *B**, common: — *M**, frequent in pastures and on borders of Winchester Reservoirs.

P. cruciata, L.

Meadows and margins of swamps. *B**, rare; damp woodpath, W. Quincy.

P. verticillata, L.

Dry soil. *B**, occasional; Pine Tree Brook valley; road, E. of Great Dome, etc.

LEGUMINOSÆ. Pulse Family.

BAPTISIA, Vent. False Indigo.

B. tinctoria, R. Br. Wild Indigo.

Dry sandy soil. *B**, common everywhere: — *M**, common in rocky woods: — *S*, frequent.

TRIFOLIUM, Tourn. Clover. Trefoil.

T. arvense, L. Rabbit-foot Clover.

Nat. from Eu. in old fields and waste places. *B** and *M*, common.

T. pratense, L. Red Clover.

Adv. from Eu. in fields and pastures. *B* and *BB**, common: — *M*, common everywhere.

T. repens, L. White Clover.

Fields and roadsides, everywhere. *B*, *M* and *BB*, common.

T. hybridum, L. Alsike Clover.

Nat. from Eu. in fields and waste places. *B*, frequent: — *M*, frequent; border of Doleful Pond; W. border of S. Reservoir, etc.: — *S**, occasional: — *B B*, occurs.

T. agrarium, L. Yellow or Hop Clover.

Nat. from Eu. by roadsides and in waste places. *B**, frequent; Hawk Hill; near Purgatory Road, etc.: — *M*, frequent in fields, open places in dry woods, and by roadsides.

MELILOTUS, Tourn. MELILOT. SWEET CLOVER.

M. officinalis, Willd. YELLOW MELILOT.

Adv. from Eu. in waste or cult. places. *B*, occasional.

MEDICAGO, Tourn. MEDICK.

M. lupulina, L. BLACK MEDICK.

Adv. from Eu. in waste places. *B*, common : — *M*, common in fields and by roadsides : — *B B**, occurs.

TEPHROSIA, Pers. HOARY PEA.

T. Virginiana, Pers. HOARY PEA. GOAT'S RUE.

Dry sandy soil and open woods. *B*, occasional ; E. slope of Great Blue Hill ; by Hoosicwhisick Pond, etc.

ROBINIA, L. LOCUST-TREE.

R. Pseudacacia, L. COMMON LOCUST.

Cult. as an ornamental tree and nat. in many places ; native South and West. *B*, clearing off Willard St. : —*M*, occasional near roads ; frequent in rocky woods, mostly as a shrub ; a tree 30 ft. high, near Melrose Reservoir : —*S* and *BB*, occurs.

DESMODIUM, Desv. TICK TREFOIL.

D. nudiflorum, DC.

Dry and rocky woods. *B**, common throughout : —*M**, frequent : —*S*, occasional.

D. acuminatum, DC.

Rich open woods. *B**, frequent ; S. slope of Great Blue Hill, etc. : —*M*, rocky open woods ; S. of Melrose Reservoir : —*S*, rare.

D. rotundifolium, DC.

Dry rocky woods. *B**, common, spreading rapidly in newly burnt ground : —*M* and *S**, dry woods.

D. cuspidatum, Torr. & Gray.

Thickets. *B**, occasional ; S. slope of Great Blue Hill ; near Hawk Hill : —*M**, rocky open woods near Black Rock : —*S*, rocky woods.

D. Dillenii, Darlingt.

Open woods. *B**, occasional; base of Bear Hill; Purgatory Road, etc.: —*M*, Cascade woods: —*S*, rare.

D. paniculatum, DC.

Copses. *B**; frequent; slope of Great Blue Hill; Hillside St.; Purgatory Road, etc.: —*M**, frequent in rocky and open woods; E. border of N. Reservoir, etc.: —*S*, rocky woods.

D. Canadense, DC.

Woods and open ground. *B**, occasional in woods, clearings, and by roadsides; valley, S. of Fox Hill, etc.

D. rigidum, DC.

Dry slopes and open places. *B*, abundant on S. slope of Great Blue Hill; rare elsewhere: —*M**, border of woods, E. side of N. Reservoir.

D. ciliare, DC.

Dry hills and fields. *B**, occasional along Hillside St.

D. Marilandicum, F. Boott.

Copses. *B**, common.

LESPEDEZA, Michx. BUSH CLOVER.

L. procumbens, Michx.

Dry sandy soil. *B**, common by roadsides and borders of woods; Rattlesnake Hill; Great Blue Hill, etc.: —*M**, occurs: —*S**, occurs in dry woods; a form of bushy upright growth, with outer stems decumbent, in woods near Turtle Pond.

L. violacea, Pers.

Dry copses. *B*, Great Blue Hill: —*M**, lower slopes of Pine Hill.

L. Virginica, Britton.† *L. reticulata*, Gray, Man. ed. 6, 141.

Open woods and roadsides. *B**, common: —*M**, occurs: —*S*, occasional.

L. Nuttallii, Darl.‡ *L. Stuvei*, Gray, Man. ed. 6, 141, in part.

† Trans. N. Y. Acad. Sci. xii. 64.
‡ Trans. N. Y. Acad. Sci. xii. 61.

Woods. *B**, occasional; Great Blue Hill, etc.: —*M*, by Winchester Reservoirs: —*S**, hills and woods near Turtle Pond.

L. intermedia, Britton.† *L. Stuvei*, var. *intermedia*, Gray, Man. ed. 6, 141.

Woods and open places. *B**, common throughout: — *M**, common in open rocky woods : — *S**, rocky woods.

L. polystachya, Michx.

Dry open places. *B**, common throughout: — *M**, frequent in open rocky woods and edges of woods: — *S**, occasional; rocky woods, etc.

L. capitata, Michx.

Dry open places. *B**, common throughout: — *M*, frequent by roadsides and borders of woods: — *S*, occasional.

VICIA, Tourn. VETCH. TARE.

V. SATIVA, L. COMMON VETCH OR TARE.

Adv. from Eu. in fields and waste places. *M*, hillside near Washington St.

V. Cracca, L.

Open ground and roadsides. *M*, low ground in open woods: —*S**, occasional; N. end of Turtle Pond, etc.

APIOS, Boerhaave. GROUND NUT. WILD BEAN.

A. tuberosa, Moench.

Low grounds, climbing over bushes. *B**, frequent; Hawk Hill; Hancock Hill, etc.: — *M*, frequent in low grounds and rocky woods: — *S*, occasional.

AMPHICARPÆA, Ell. HOG PEANUT.

A. monoica, Nutt.

Rich woods and shaded roadsides, twining. *B**, common throughout: —*M*, frequent : — *S**, rocky woods: — *B B**, occurs.

† Trans. N. Y. Acad. Sci. xii. 63.

ROSACEÆ. Rose Family.

PRUNUS, Tourn. Plum. Cherry.

P. Persica, Sieb. & Zucc. Peach.

Cult. from Persia. *B**, several trees on house-site, Hillside St.

P. domestica, L. Common Plum.

Cult. probably from Asia. *B B*, a well-established patch of Damson Plums, near wall below The Falls.

P. pumila, L. Dwarf Cherry. Sand Cherry.

Rocky and sandy places. *B**, occasional on or near hilltops in the seams of ledges: —*M*, occasional.

P. Cerasus, L. Garden Cherry.

Cult. from Eu. *B*, occasional on house-sites and by roadsides: —*M*, one tree, border of wood-road, Melrose.

P. avium, L. Bird Cherry.

Cult. from Eu. and often escaped into woods. *B**, single tree on house-site, Hillside St.: —*M*, occasional.

P. Pennsylvanica, L. f. Wild Red Cherry.

Rocky soil, woods, and thickets. *B**, common in dry ledgy places and on hill-tops, often on fire-burned slopes where it is soon shaded out by stronger plants: —*M**, frequent in dry, rocky, and upland woods: —*S*, common on Milkweed Hill: — *B B*, occasional.

P. Virginiana, L. Choke Cherry.

Dry rocky places, borders of woods and roadsides. *B**, frequent: —*M**, common: —*B B*, occurs.

P. serotina, Ehrh. Wild Black Cherry.

Woods and open places. *B*, frequent, usually on ledges: — *M**, common on borders of woods, and frequent by roadsides.

SPIRÆA, L. Spiræa. Meadow Sweet.

S. tomentosa, L. Hardhack. Steeple Bush.

Low ground. *B** and *M*, common in meadows and pastures,

but shaded out by the growth about it: — *S*, occasional: — *B B*, frequent.

S. salicifolia, L. COMMON MEADOW SWEET.

Low ground. *B**, *M*, *S* and *B B*, very common in pastures, meadows, and by roadsides, but shaded out by the growth about it.

S. JAPONICA, L.

Cult. from Japan and China. *B*, old garden, Park's place, Hawk Hill.

S. PRUNIFOLIA, Sieb.

Cult. from Japan. *B*, old garden, Park's place, Hawk Hill.

S. SORBIFOLIA, L.

Cult. from Siberia. *B*, house-site, N. of Blueberry Swamp.

S. FILIPENDULA, L. DROPWORT.

Cult. from Eu. *B*, old garden, Park's place, Hawk Hill.

RUBUS, Tourn. RASPBERRY. BLACKBERRY.

R. triflorus, Richards. DWARF OR WOOD RASPBERRY.

Ascending or trailing in damp woods and swamps. *B* and *M*, frequent.

R. strigosus, Michx. WILD RED RASPBERRY.

Open thickets, old fields, clearings, and roadsides. *B**, *M*, *S* and *B B**, common on edges of wet ground and on hillsides and roadsides.

R. occidentalis, L. THIMBLEBERRY.

Borders of fields and thickets, especially where ground has been burnt over. *B**, frequent in open places among rocks, and in low thickets: — *M**, common on rocky banks and in open woody places: — *B B**, occurs.

R. villosus, Alt. HIGH BLACKBERRY.

Borders of thickets, fence-rows, and roadsides. In all the Reservations, in old pastures, on edges of wet runs and borders of meadows, not persisting long in shade and partly grown-up pastures: —*B* and *M*, common: — *S**, frequent: — *B B**, occasional.

R. Canadensis, L. Low Blackberry. Dewberry.

Rocky and sandy soil, long-trailing. B, common: — M, common in rocky woods and low grounds: — S^* and BB^*, occurs.

R. hispidus, L. Running Swamp Blackberry.

Low grounds and sandy places. In all the Reservations, in wet meadows, moist fields, dry pastures, and under light shade; spreads rapidly, but is seriously injured by fire: — M, common: — B^*, S^* and BB, frequent.

GEUM, L. Avens.

G. album, Gmelin.

Thickets and borders of woods. B and BB, occurs: — M^*, frequent.

G. Virginianum, L.

Thickets and borders of woods. M, frequent: — BB, shady meadows.

G. rivale, L. Water Avens.

Meadows and bogs. M, occasional: — BB, common by the brook.

FRAGARIA, Tourn. Strawberry.

F. Virginiana, Mill.

Moist woodlands, fields, and light shady places. B and M, common: — BB, occurs.

F. vesca, L.

Moist woodlands, fields, and light shady places. M^*, Bear Hill; low woods off Ravine Road: — BB^*, occurs.

POTENTILLA, L. Cinquefoil. Five-finger.

P. arguta, Pursh.

Sunny rocky slopes. B, rare; ledges on Great Blue Hill.

P. Norvegica, L.

Moist or wet, open or partly shaded soil. B and M, common: — S, occurs.

P. argentea, L. Silvery Cinquefoil.

Thin soil in paths, pastures and on rocks in woods, in sun or light shade. *B**, *M*, *S** and *B B*, common.

P. palustris, Scop. Marsh Five-finger.

Swamps and wet places. *S**, near Turtle Pond.

P. Canadensis, L. Common Cinquefoil. Five-finger.

Fields, meadows, roadsides, and open woods. *B**, *M**, *S** and *B B**, common.

AGRIMONIA, Tourn. Agrimony.

A. Eupatoria, L. Common Agrimony.

Borders of woods, pastures, and roadsides. *B**, frequent:— *M**, common : — *B B**, occurs.

ROSA, Tourn. Rose.

R. Carolina, L. Swamp Rose.

Moist low thickets, and borders of swamps and streams. *M**, common : — *S* and *B B**, occurs.

R. lucida, Ehrh. Common Wild Rose.

Low thickets, fields and roadsides. *B*, *M*, *S** and *B B*, frequent on the edges of meadows, in pastures, and on ledges.

R. humilis, Marsh.

Dry soil and rocky slopes. *B**, occasional; E. slope of Hancock Hill, etc. : — *M*, Bear Hill : — *S**, occasional.

R. rubiginosa, L. *R. micrantha,* Smith. Sweet Brier. Eglantine.

Int. from Eu. into pastures, thickets, and roadsides. *B**, frequent on Hawk Hill, etc. : —*M*, Pine Hill ; woods on Bear Hill, etc. : —*BB*, occurs.

R. cinnamomea, L. Cinnamon Rose.

Cult. from Eu. and occasionally escaped. *B**, house-site on Hillside St.

R. canina, L. Dog Rose.

Int. from Eu. and occasionally escaped. *B*, old garden, Park's place. Hawk Hill.

R. Manettii, Hort.

An Italian rose of unknown origin, used as stocks on which to bud garden roses. *B*, growing and spreading somewhat in old garden, Park's place, Hawk Hill.

PYRUS, L. Pear. Apple.

P. communis, L. Common Pear.

Cult. from Eu. and occasionally self-sown in pastures, by roadsides, etc. *B*, *M*, *S* and *BB**, house-sites, fields, and open woods.

P. Malus, L. Common Apple.

Cult. from Eu. and frequently self-sown in pastures, by roadsides, etc. *B*, *M*, *S* and *BB*, house-sites, fields, woods, pastures, old orchards and cellars.

P. arbutifolia, L. f. Chokeberry.

Damp thickets, swamps, fields, roadsides, etc. *B**, frequent in wet and dry soil and on exposed hill-tops, not persisting long in shade:—*M** and *S**, common.

Var. melanocarpa, Hook.

Same habitat as the species. *B** and *M*, common.

P. Aucuparia, Gaertn. European Mountain Ash.

Cult. from Eu. and occasionally escaped. *B**, several escaped trees opposite Old Houghton Place:—*M**, frequent as a shrub in woods; several slender trees on border of wood:—*BB*, a small tree.

P. Cydonia, L. Common Quince.

Cult. from Eu. and occasionally escaped. *M*, a few trees in old orchard off Pond St.

CRATÆGUS, L. Hawthorn. Thorn.

C. Oxyacantha, L. English Hawthorn.

Cult. from Eu. and occasionally escaped. *BB**, frequent under Waverly Oaks.

AMELANCHIER, Medic. SHADBUSH. JUNE BERRY. SUGAR PLUM.

A. Canadensis, Torr. & Gray. SHADBUSH. SERVICE BERRY.

Dry or wet places, in sun or shade. In all the Reservations from the edges of lowlands to the tops of the hills :—*B*, frequent : —*M*, common : —*S* and *BB*, occasional.

Var. (?) **oblongifolia,** Torr. & Gray.

Low ground, swampy open woods, and hill-tops. *B**, frequent : —*M**, common : —*S*, occasional.

SAXIFRAGACEÆ. SAXIFRAGE FAMILY.

SAXIFRAGA, L. SAXIFRAGE.

S. Virginiensis, Michx. EARLY SAXIFRAGE.

Dry hillsides, open or shaded ledges, and brooksides. *B*, frequent: —*M** and *BB*, common.

S. Pennsylvanica, L. SWAMP SAXIFRAGE.

Meadows and swamps. *B* and *M*, frequent : —*BB*, common.

CHRYSOSPLENIUM, Tourn. GOLDEN SAXIFRAGE.

C. Americanum, Schwein.

Shaded wet grounds, and by springs and brooks. *B**, rare : —*M**, frequent : —*S**, occurs : —*BB**, common.

DEUTZIA, Thunb.

D. GRACILIS, Sieb. & Zucc.

Cult. from Japan. *B*, old garden, Park's place, Hawk Hill.

D. SCABRA, Thunb.

Cult. from China and Japan. *B*, old garden, Park's place, Hawk Hill.

HYDRANGEA, Gronov.

H. PANICULATA, Sieb. HARDY HYDRANGEA.

Cult. from Japan. *B*, a large-flowered variety in old garden, Park's place, Hawk Hill.

PHILADELPHUS, L. MOCK ORANGE. SYRINGA.

P. CORONARIUS, L.
Cult. from S. Eu. B*, old garden, Park's place, Hawk Hill; roadside opp. Old Houghton Place.

P. GORDONIANUS, Lindl.
Cult. from Cal. and Oregon. B, old garden, Park's place, Hawk Hill.

RIBES, L. GOOSEBERRY. CURRANT.

R. oxyacanthoides, L. WILD GOOSEBERRY.
Rocky moist slopes and open woods. B*, common: — M, occasional.

R. RUBRUM, L. GARDEN CURRANT.
Cult. from Eu. B*, about old cellar, Hawk Hill; house-site, Hillside St.

Var. **subglandulosum**, Maxim. RED CURRANT.
Damp woods, and bogs. M*, rare; near Ravine Road.

R. NIGRUM, L. GARDEN BLACK CURRANT.
Cult. from Eu. B, persisting in old garden, house-site, E. of Hillside St.

CRASSULACEÆ. ORPINE FAMILY.

PENTHORUM, Gronov. DITCH STONECROP.

P. sedoides, L.
Bogs, and borders of swamps and streams. B*, M and BB*, occurs sparingly.

SEDUM, Tourn. STONECROP. ORPINE.

S. ACRE, L. MOSSY STONECROP.
Cult. from Eu. and escaped to rocky places. B, occurs.

S. TELEPHIUM, L. LIVE-FOR-EVER.
Cult. from Eu. in old gardens and escaped to roadsides, etc. B*, house-site, Hillside St. : —M, occurs.

DROSERACEÆ. Sundew Family.

DROSERA, L. Sundew.

D. rotundifolia, L. Round-leaved Sundew.

Sphagnous bogs, meadows, sandy shores of ponds, etc. B^*, frequent: — S^*, edge of Turtle Pond.

D. intermedia, Hayne, var. Americana, DC.

Bogs, borders of ponds, etc. B and M^*, frequent: — S^*, edge of Turtle Pond.

HAMAMELIDEÆ. Witch Hazel Family.

HAMAMELIS, L. Witch Hazel.

H. Virginiana, L.

Damp or dry woods. B^*, common in moist shady runs, at the base of rocky slopes, and on hillsides: — M^*, common on damp rocky hillsides and in low woods: — S^* and BB^*, occurs.

HALORAGEÆ. Water Milfoil Family.

MYRIOPHYLLUM, Vaill. Water Milfoil.

M. ambiguum, Nutt., var. capillaceum, Torr & Gray.

Ponds and slow streams. M, frequent along the S. and W. shores of Spot Pond.

PROSERPINACA, L. Mermaid Weed.

P. palustris, L.

Swamps, wet borders of ponds, and meadows. B^* and M^*, occasional.

CALLITRICHE, L. Water Starwort.

C. verna, L.

Ponds and still water. B, small form in mud, entirely emersed: — M, common in ditches and stagnant water.

C. heterophylla, Pursh.

Ponds and stagnant water. M^*, ditch by Spot Pond; form entirely immersed, Stoneham: — S^*, swamp, S. W. of Office.

MELASTOMACEÆ. Melastoma Family.

RHEXIA, L. Meadow Beauty.

R. Virginica, L.

Open, wet, sandy or gravelly soil and open meadows. *B**, common round Hoosicwhisick Pond.

LYTHRACEÆ. Loosestrife Family.

LYTHRUM, L. Loosestrife.

L. Hyssopifolia, L.

Boggy and wet meadows. *B**, occurs.

DECODON, Gmel. Swamp Loosestrife.

D. verticillatus, Ell.

Bogs and wet places. *B**, frequent; by Hoosicwhisick Pond; Cedar Swamp, etc. : — *M*, by Doleful Pond : — *S*, by Turtle Pond.

ONAGRACEÆ. Evening Primrose Family.

LUDWIGIA, L. False Loosestrife.

L. palustris, L. Water Purslane.

Bogs, ditches, and wet ground. *B**, frequent : — *M**, common.

EPILOBIUM, L. Willow Herb.

E. angustifolium, L. Fireweed.

Low ground, rocky places, and especially where forests have been cleared, and ground burnt over. *B** and *M**, common : — *S**, frequent.

E. lineare, Muhl.

Bogs and wet ground. *B**, Cedar Swamp; by Hoosicwhisick Pond : — *M*, common.

E. coloratum, Muhl.

In all wet places. *B**, occasional; near the Old Glover Place; near Hoosicwhisick Pond, etc.

E. adenocaulon, Haussk.

In all wet places. *B**, Pine Tree Brook meadow ; pool, S. E. of Great Dome : — *M**, frequent.

ŒNOTHERA, L. Evening Primrose.

Œ. biennis, L. Common Evening Primrose.

Fields, roadsides, and open grounds. *B* and *M*, common.

Œ. pumila, L.

Dry soil in open places. *B*, occurs : — *M*, common : — *S**, occasional.

CIRCÆA, Tourn. Enchanter's Nightshade.

C. Lutetiana, L.

Moist shady places. *B**, occasional : — *M*, occasional in damp open woods : — *BB*, rare.

CUCURBITACEÆ. Gourd Family.

ECHINOCYSTIS, Torr. & Gray. Wild Balsam Apple.

E. lobata, Torr. & Gray.

Low grounds and waste places, native West; extensively cult. for arbors, and run wild. *M*, roadside : — *S**, house-yard.

FICOIDEÆ.

MOLLUGO, L. Indian Chickweed.

M. verticillata, L. Carpet Weed.

Roadsides and about cult. grounds; from the tropics. *B**, common.

UMBELLIFERÆ. Parsley Family.

DAUCUS, Tourn. Carrot.

D. Carota, L.

Cult. from Eu. and run wild in fields and by roadsides. *M*, common.

PASTINACA, L. Parsnip.

P. sativa, L.

Cult. from Eu. and run wild in old fields and waste places. *B**, common : — *M* and *S*, occasional.

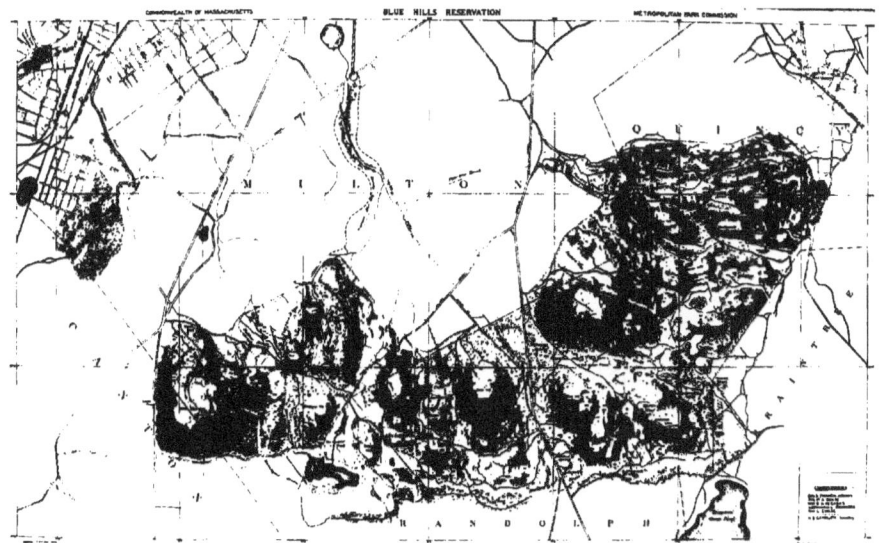

O. Claytoniana, L.

Low ground, with *Onoclea sensibilis*. *B* * and *M*, common: —*S*, woods near Turtle Pond.

O. cinnamomea, L. CINNAMON FERN.

Swamps and low ground; abnormal forms frequent. *B* * and *M*, common: — *S*, damp woods.

OPHIOGLOSSACEÆ. ADDER'S TONGUE FAMILY.†

BOTRYCHIUM, Swz. GRAPE FERN. MOONWORT.

B. lanceolatum, Angstroem.

Collected some years ago in the Fells, and should therefore be found now; may be looked for with the following species in low spongy woodlands.

B. matricariæfolium, Braun.

Collected once in the Fells, but no specimen has been received.

B. ternatum, Swz.

Occasional in deep woods, on borders of swamps. *B* *, rare; dense woods: — *M*, rare. As represented by the var. *intermedium*, D. C. Eaton, doubtless in all or most of the Reservations. The common forms by which the species are represented with us are the following varieties.

Var. **obliquum,** Milde.

Frequent in old pastures and on moist hillsides. *B* *, occasional: — *M* *, meadow, S. of Pine Hill.

Var. **dissectum,** Milde.

Frequent in old pastures and on moist hillsides. *B* *, rare; Barberry Bush Spring, etc.

B. Virginianum, Swz. RATTLESNAKE FERN.

Rich woods. *B*, rare; Cedar Swamp: — *M* *, occasional in the E. part of the Reservation; boggy woods, end of S. Reservoir: — *BB* *, woods below lower dam.

† *Ophioglossum vulgatum*, L., Adder's Tongue, should be found in some of the old pasture lands in the Reservations.

LYCOPODIACEÆ. Club Moss Family.

LYCOPODIUM, L. Club Moss.

L. lucidulum, Michx.
Occasional in damp woods. *B**, occurs.

L. inundatum, L.
Sandy bogs; generally represented in this section by the var. *Bigelovii*, Tuck. *S*, border of Turtle Pond.

L. obscurum, L., var. **dendroideum,** D. C. Eaton.
Dry woods. *B**, occasional: — *M*, common.

L. clavatum, L. Club Moss.
Dry woods. *B*, reported as occasional: — *M*, collected in years past by G. E. Davenport.

L. complanatum, L. Ground Pine.
Frequent in dry woods and thickets. *B**, occurs: — *M**, slope of Bear Hill.

SELAGINELLACEÆ.

An order comprising a small group of terrestrial plants resembling Lycopods, and a group of aquatic or semi-aquatic plants growing in mud or water.

SELAGINELLA, Beauv.

S. rupestris, Spring.
Dry exposed rocks. *B*, top of Great Blue Hill; by Monatiquot Stream: — *M*, Cascade ledges; Bear Hill.

S. apus, Spring.
Low ground in shady places. *M*, collected by a brook on the E. side of Bear Hill about 1880, but not found since.

ISOETES, L. Quillwort.

I. echinospora, Durieu, var. **Braunii,** Engelm.
Ponds and shallow water; the type not represented in this country. *M**, shore of Spot Pond.

I. Engelmanni, Braun.

Ponds and shallow water; immersed. *B **, stagnant pool, W. Quincy; pool in Pine Tree Brook.

NOTE. Two more species and as many varieties of *Isoetes* should be found within our limits. *Marsilia quadrifolia*, L., with four-parted leaves like an *Oxalis* should also be found in some of the ponds and streams as it has been distributed many times from the Cambridge Botanic Garden. It grew abundantly with *Trapa natans*, L. some years ago in a pond near Highland Ave., Medford.

CLASS II. BRYOPHYTA.†

DIVISION I. MUSCI. MOSSES.

(By Edward L. Rand.)

The list of mosses here given is based mainly on the valuable collections of Messrs. Edwin and Charles E. Faxon. As these collections, however, were made mostly in the Blue Hills and Stony Brook Reservations, the moss flora of the other Reservations is most imperfectly represented in the list,— in fact only two species have been reported from the Beaver Brook Reservation.

The list of Sphagnaceæ is arranged in accordance with the articles of Dr. Carl Warnstorf in Vol. XV of Coulter's Botanical Gazette. Synonyms, however, have been given when possible to facilitate reference to Lesquereux and James' " Mosses of North America." The lists of Andreæaceæ and Bryaceæ are arranged in accordance with the Manual of Lesquereux and James just mentioned, in the absence of a standard work of ready reference of a more recent date.

ORDER I. **SPHAGNACEÆ.** PEAT MOSSES.

SPHAGNUM, L. PEAT MOSS.

S. Girgensohnii, Russ. *S. strictum*, Lindb.

B, rare.

† The Hepaticæ have been omitted, owing to the illness of Miss Cora H. Clark who was to have prepared the list.

S. tenellum, von Klinggraef.†
  B and *S*, rare.

S. acutifolium, Russ. & Warnst. *S. acutifolium*, Ehrh. in part.
  B and *S*, common.

S. cuspidatum, Russ. & Warnst.
  B and *S*, common : — *M*, frequent.

S. recurvum, Russ. & Warnst. *S. intermedium*, Hoffm.
  B and *S*, frequent.

S. subsecundum, Nees.
  B and *S*, frequent.

S. imbricatum, Russ. *S. Austini*, Sulliv.
  B and *S*, common : — *M*, frequent.

S. cymbifolium, Ehrh.
  B and *S*, common.

S. medium, Limpr.
  B and *S*, common.

ORDER II. **ANDREÆACEÆ.** SCHIZOCARPOUS MOSSES.

ANDREÆA, Ehrh.

A. rupestris, Turner.
  On rocks. *B* and *S*, rare.

A. petrophila, Ehrh.
  On rocks. *B*, rare.

ORDER III. **BRYACEÆ.** TRUE MOSSES.

WEISIA, Hedw.

W. viridula, Brid.
  On ground. *B* and *S*, common.

† Bot. Gaz. xv. 135.

DICRANELLA, Schimp.

D. heteromalla, Schimp.
On rocks, ground, base of trees, etc. B and S, common.

DICRANUM, Hedw.

D. montanum, Hedw.
On decaying trees. B and S, common.

D. flagellare, Hedw.
On decaying trees. B and S, common.

D. fulvum, Hook.
On rocks. B and S, common.

D. fuscescens, Turn.
On rocks and decaying trees. B and S, common.

D. scoparium, Hedw.
On ground, etc. B and S, common.

D. Schraderi, Web. & Mohr.
Moist ground. B and S, common.

D. spurium, Hedw.
Ledges. B and S, rare.

D. undulatum, Turn.
Moist ground. B and S, common.

FISSIDENS, Hedw.

F. osmundoides, Hedw.
Moist ground, etc. M, rare.

F. taxifolius, Hedw.
Shaded ground. M and S, occasional.

F. adiantoides, Hedw.
Wet ground and rocks. M and S, occasional.

CONOMITRIUM, Mont.

C. Julianum, Mont.
On stones and branches in moist ground. BB, rare.

LEUCOBRYUM, Hampe.

L. vulgare, Hampe.
On ground. *B*, *M* and *S*, common.

CERATODON, Brid.

C. purpureus, Brid.
On rocks and ground. *B* and *S*, common.

POTTIA, Ehrh.

P. truncata, Fuern.
On ground. *B* and *S*, common.

LEPTOTRICHUM, Hampe.

L. tortile, Muell.
On ground. *S*, occasional.

L. vaginans, Lesq. & James.
On ground. *B* and *S*, occasional.

L. pallidum, Hampe.
On ground in woods. *B* and *S*, frequent.

BARBULA, Hedw.

B. unguiculata, Hedw.
On ground and rocks. *S*, occasional.

B. cæspitosa, Schwaegr.
On roots of trees. *B*, occasional.

B. papillosa, Muell.
On tree trunks. *B*, rare.

GRIMMIA, Ehrh.

G. conferta, Funck.
On rocks. *B* and *S*, common.

G. apocarpa, Hedw.
On rocks. *B* and *S*, common.

G. Olneyi, Sulliv.
On rocks. *B* and *S*, rare.

G. Pennsylvanica, Schwaegr.
On rocks. *B* and *S*, common.

RACOMITRIUM, Brid.

R. aciculare, Brid.
On wet rocks. *B* and *S*, occasional.

R. Sudeticum, Bruch & Schimp.
On rocks. *B*, *M* and *S*, occasional.

HEDWIGIA, Ehrh.

H. ciliata, Ehrh.
On rocks. *B*, *M* and *S*, common.

DRUMMONDIA, Hook.

D. clavellata, Hook.
On trees. *B*, rare.

ULOTA, Mohr.

U. Ludwigii, Brid.
On trees. *B* and *S*, common.

U. crispula, Brid.
On trees. *B* and *S*, rare.

U. Hutchinsiæ, Schimp.
On rocks. *B*, *M* and *S*, common.

ORTHOTRICHUM, Hedw.

O. cupulatum, Hoffm., var. minus, Sulliv.
On rocks. *S*, rare.

O. sordidum, Sulliv. & Lesq.
On trees. *B* and *S*, common.

O. strangulatum, Beauv.
On trees. *B* and *S*, occasional.

TETRAPHIS, Hedw.

T. pellucida, Hedw.
On decayed trees. *B* and *S*, common.

PHYSCOMITRIUM, Brid.

P. pyriforme, Brid.
On ground. *B* and *S*, common.

FUNARIA, Schreb.

F. hygrometrica, Sibth.
On ground, especially on burnt soil. *B*, *M* and *S*, common.

BARTRAMIA, Hedw.

B. pomiformis, Hedw.
On rocks and shady banks. *B*, *M* and *S*, occasional.

PHILONOTIS, Brid.

P. fontana, Brid.
Wet places and moist rocks. *B*, *M* and *S*, occasional.

WEBERA, Hedw.

W. nutans, Hedw.
On ground. *B* and *S*, common.

BRYUM, L.

B. bimum, Schreb.
On decayed trees, etc. *B* and *S*, common.

B. argenteum, L.
On ground, etc. *B* and *S*, common.

B. cæspiticium, L.
On ground, etc. *B* and *S*, common.

B. capillare, L.
On rich soil. *B* and *S*, common.

B. pseudotriquetrum, Schwaegr.
On wet ground and rocks. *B* and *S*, common.

B. roseum, Schreb.
Shaded ground. *B* and *S*, rare.

MNIUM, L.

M. cuspidatum, Hedw.
On ground, shaded places. *B* and *S*, common.

M. affine, Bland.
On ground, shaded places. *B* and *S*, common.

M. hornum, L.
On ground and rocks in shaded places. *B* and *S*, common.

M. punctatum, Hedw.
Damp ground. *B* and *S*, occasional.

AULACOMNIUM, Schwaegr.

A. palustre, Schwaegr.
Wet places. *B* and *S*, common.

A. heterostichum, Bruch & Schimp.
Shaded places. *B* and *S*, common.

ATRICHUM, Beauv.

A. undulatum, Beauv.
Damp ground. *B* and *S*, common.

A. angustatum, Bruch & Schimp.
On ground. *B* and *S*, common.

POGONATUM, Beauv.

P. brevicaule, Beauv.
On clay soil. *B* and *S*, occasional.

POLYTRICHUM, L.

P. Ohioense, Ren. & Card.† *P. formosum*, Lesq. & J., Mosses N. A., 264, in part.
On ground in woods. *B* and *S*, occasional.

P. piliferum, Schreb.
Dry ground. *B* and *S*, common.

P. juniperinum, Willd.
Dry ground. *B* and *S*, common.

† See Bot. Gaz. xiii. 199.

P. commune, L.
Dry or moist ground, woods and open places. *B, M, S* and *B B*, common.

DIPHYSCIUM, Mohr.

D. foliosum, Mohr.
Shaded clayey or loamy banks. *B* and *S*, common.

BUXBAUMIA, Hall.

B. aphylla, L.
On ground. *B* and *S*, occasional.

FONTINALIS, Dill.

F. antipyretica, L., var. **gigantea,** Sulliv.
Streams. *S*, occasional.

F. Novæ-Angliæ, Sulliv.
Streams. *S*, common.

F. Lescurii, Sulliv.
Streams. *S*, common.

DICHELYMA, Myrin.

D. pallescens, Bruch & Schimp.
Pools. *S*, common.

LEUCODON, Schwaegr.

L. julaceus, Sulliv.
On trees. *B* and *S*, common.

THELIA, Sulliv.

T. hirtella, Sulliv.
At base of trees. *B* and *S*, common.

T. asprella, Sulliv.
At base of trees. *B* and *S*, common.

LESKEA, Hedw.

L. obscura, Hedw.
Roots and at base of trees. *B* and *S*, common.

ANOMODON, Hook. & Tayl.

A. rostratus, Schimp.
On roots of trees. *B* and *S*, common.

A. attenuatus, Hueben.
On rocks and roots of trees along streams. *B*, common.

A. obtusifolius, Bruch & Schimp.
On trunks of trees. *S*, rare.

PYLAISIA, Bruch & Schimp.

P. intricata, Bruch & Schimp.
On trees and old logs. *B* and *S*, common.

P. velutina, Bruch & Schimp.
On trees. *S*, common.

CYLINDROTHECIUM, Bruch & Schimp.

C. seductrix, Sulliv.
On logs, moist shaded places. *B* and *S*, common.

CLIMACIUM, Web. & Mohr.

C. Americanum, Brid.
Damp shaded places. *B* and *S*, common :—*M*, occasional.

HYPNUM, Dill.

H. recognitum, Hedw.
On ground, rocks, etc. *B* and *S*, common.

H. delicatulum, L.
On ground. *B* and *S*, common.

H. paludosum, Sulliv.
Wet ground. *S*, rare.

H. lætum, Brid.
Shaded places on roots, logs, etc. S, rare.

H. salebrosum, Hoffm.
On ground, stones, etc. B and S, common.

H. velutinum, L.
On ground, shaded places. B and S, common.

H. rutabulum, L.
On ground, roots, and stones. B and S, common.

H. Novæ-Angliæ, Sulliv. & Lesq.
On ground. B and S, common.

H. populeum, Hedw.
Usually on rocks. B and S, common.

H. plumosum, Swartz.
On moist rocks. B and S, common.

H. strigosum, Hoffm.
On ground. B and S, common.

H. Boscii, Schwaegr.
On ground. S, rare.

H. serrulatum, Hedw.
On ground. B and S, common.

H. rusciforme, Weis.
On stones in water. S, rare.

H. Alleghaniense, Muell.
Shaded banks. M and S, occasional.

H. denticulatum, L.
On decaying trees. M, rare.

H. radicale, Beauv.
On decaying wood, roots, etc. S, common.

H. orthocladon, Beauv.
On the ground, stones, etc., in damp places. M, rare.

H. adnatum, Hedw.
On stones or base of trees. S, common.

H. riparium, L.
On stones, roots, etc., in still water. *B* and *S*, common.

H. fluitans, L.
In still water. *B* and *S*, common.

H. Crista-castrensis, L.
Pine woods, on ground. *S*, rare.

H. reptile, Michx.
On trees. *S*, common.

H. imponens, Hedw.
On decaying trees, etc. *B*, common.

H. cupressiforme, L.
On trees, stones, etc. *B* and *S*, common.

H. curvifolium, Hedw.
On decaying logs. *S*, rare.

H. Haldanianum, Grev.
On damp ground and decaying trees. *B* and *S*, common.

H. cordifolium, Hedw.
Wet places. *B* and *S*, common.

H. Schreberi, Willd.
On ground in shaded places. *B*, *M* and *S*, common.

H. splendens, Hedw.
On ground in woods. *B*, rare.

H. triquetrum, L.
On ground in woods. *B*, rare.

CLASS III. THALLOPHYTA.

DIVISION I. CHARACEÆ.
(By J. W. Blankinship.)

NITELLA, Ag.

N. flexilis, Ag.
B B, in the brook.

N. mucronata, A. Br.
  B B, in the brook.

CHARA, L.
C. fragilis, Desv.
  B B, upper pond.

Division II. ALGÆ. †
(By F. S. Collins.)

MYXOPHYCEÆ.

CLATHROCYSTIS, Henfrey.
C. æruginosa, Henfrey.
  M, forming a floating scum on Middle Reservoir.

CŒLOSPHÆRIUM, Naeg.
C. Kuetzingianum, Naeg.
  M, scattered, or as a scum, on Spot Pond.

GLŒOCAPSA, Naeg.
G. Magma (Bréb.), Kuetz.
  M, forming a dark purplish slimy coating on wet rocks throughout; not uncommon, but usually in small quantity.

G. polydermatica, Kuetz.
  M, on dripping rocks, Cascade. Determination somewhat uncertain, as with many species of this genus.

SCHIZOTHRIX, Kuetz.
S. lacustris, A. Br., var. **cæspitosa,** Gomont.
  M, on stones along the margin of Spot Pond.

S. Muelleri, Naeg.
  M, forming a black coating on wet rocks; easily stripped off in sheets of considerable size.

† The specimens on which this list is founded are in the herbarium of Mr. F. S. Collins.

PLECTONEMA, Thuret.

P. tenue, Thuret.
M, Spot Pond.

SYMPLOCA, Kuetz.

S. muralis, Kuetz.
M, forming minute green plush-like patches on ground, near Black Rock.

LYNGBYA, Ag.

L. ochracea, Thuret.
M, Cascade : — *B B*, on rocks in stream. Probably everywhere common.

PHORMIDIUM, Kuetz.

P. Corium (Ag.), Gomont.
M, on cliff.

P. uncinatum, Gomont.
M, running brook near Elm St., Medford.

OSCILLATORIA, Vauch.

O. amœna (Kuetz.), Gomont.
B B, on rocks and trunks of trees.

O. amphibia, Ag.
B B, with the preceding species.

O. tenuis, Ag.
B B, with the preceding species.

O. terebriformis, Ag.
B B, with the preceding species.

CALOTHRIX, Ag.

C. fusca (Kuetz.), Born. & Flah.
M, among various gelatinous algæ on rocks at Cascade.

C. parietina (Naeg.), Thuret.
M, forming minute tufts on rocks near Bear's Den.

HAPALOSIPHON, Naeg.
H. pumilus (Kuetz.), Kirchner.
M, on under side of *Nuphar* leaves, Spot Pond and Shiner Pool.

STIGONEMA, Ag.
S. hormoides (Kuetz.), Born. & Flah.
M, in gelatinous masses on rocks, Cascade.

S. panniforme (Kuetz.), Born. & Flah.
M, wet rock.

S. mamillosum, Ag.
M, Cascade; on pebbles at margin of Spot Pond.

HASSALLIA, Berk.
H. byssoidea (Berk.), Hassall.
M, among other algæ, Cascade.

TOLYPOTHRIX, Kuetz.
T. tenuis, Kuetz.
M, on mosses and various small plants, Spot Pond.

NOSTOC, Vauch.
N. sphæricum, Vauch.
M, minute blackish or greenish rounded masses on wet rocks, Cascade.

ANABÆNA, Bory.
A. catenula (Kuetz.), Born. & Flah.
M, on dead leaves, etc., in swamp near Bear's Den path.

A. oscillarioides, Bory.
M, with the preceding species.

CONJUGATÆ.

ZYGNEMA, Kuetz.
Z. cruciatum (Vauch.), Ag.
M, in pond.

Z. insigne, Kuetz.
  M, Cascade; near Spot Pond.
Z. stellinum, Ag.
  M, swamp near Bear's Den path.

SPIROGYRA, Link.

S. bellis (Hass.), Cleve.
  BB, in pond.
S. calospora, Cleve.
  M, rather common.
S. catenæformis (Hass.), Kuetz.
  M, Spot Pond brook, Virginia Wood.
S. crassa, Kuetz.
  BB, in pond and brook.
S. Grevilleana (Hass.), Kuetz.
  M, swamps near Highland Ave., Medford.
S. inflata (Vauch.), Rab.
  M, swamp near Bear's Den.
S. insignis (Hass.), Kuetz., var. **Hantzschii** (Rab.), Petit.
  M, brook, Virginia Wood : — *BB*, in pond.
S. orthospira (Naeg.), Kuetz.
  M, brook, Virginia Wood.
S. quadrata (Hass.), Petit.
  M, Melrose.
S. Spreeiana, Rab.
  M, Virginia Wood; Pine Hill.
S. varians (Hass.), Kuetz.
  BB, in pond.

ZYGOGONIUM, Kuetz.

Z. ericetorum, De Bary.
  M, Pine Hill.

MOUGEOTIA, Ag.

M. mirabilis (A. Br.), Wittr.

M, swamp near Bear's Den, fruiting plentifully in June, 1893.

M. nummuloides, Hass.
M, swamp near Bear's Den.

M. scalaris, Hass.
M, swamp near Elm St., Medford.

M. quadrangulata, Hass.
M, swamp near Bear's Den.

CHLOROPHYCEÆ.

VOLVOX, Ehren.

V. globator, Ehren.
M, Shiner Pool.

TETRASPORA, Ag.

T. lubrica, Ag.
M, very common in spring in running water.

OPHIOCYTIUM, Naeg.

O. parvulum (Perty), A. Br.
M, sparsely distributed among various algæ.

PEDIASTRUM, Meyer.

P. Boryanum (Turp.), Menegh.
M, among various algæ, Hemlock Pool.

ULOTHRIX, Kuetz.

U. zonata (Web. & Mohr), Kuetz.
M, Spot Pond brook in spring. Other species of *Ulothrix*, also species of *Conferva*, are common in the Fells, but specific determinations are uncertain.

CHÆTOPHORA, Schrank.

C. Cornu-Damæ (Roth), Ag.
M, common.

C. pisiformis (Roth), Ag.

M, common in brooks on sticks, mosses, etc.; most abundant in spring.

DRAPARNALDIA, Ag.

D. glomerata (Vauch.), Ag.

M, very common in running water in spring.

STIGEOCLONIUM, Kuetz.

S. tenue (Ag.), Kuetz.

M, common in the same places as the preceding species.

APHANOCHÆTE, A. Br.

A. repens, A. Br.

M, on filamentous algæ, Pine Hill and Hemlock Pool: —*BB*, on algæ in brook.

TRENTEPOHLIA, Mart.

T. umbrina (Kuetz.), Born.

M, common on bark of trees.

ŒDOGONIUM, Link.

Œ. acrosporum, De Bary.

M, swamp near Bear's Den.

Œ. Borisianum (Le Cl.), Wittr.

B B, in pond.

Œ. crassiusculum, Wittr.

B B, with the preceding species.

Œ. fragile, Wittr.

M, swamp near Bear's Den.

Œ. Landsboroughi (Hass.), Wittr.

M, Hemlock Pool: — *B B*, in pond.

BULBOCHÆTE, Ag.

B. setigera (Roth), Ag.

M, Hemlock Pool.

COLEOCHÆTE, Bréb.

C. orbicularis, Prings.

M, on under surface of *Nuphar* and *Nymphœa* leaves, Spot Pond.

VAUCHERIA, DC.

V. geminata (Vauch.), DC., var. **racemosa,** Walz.

M, dense floating masses, swamp near Highland Ave., Medford.

V. sessilis (Vauch.), DC.

M, in moist places, but not under water, Cascade and Virginia Wood.

V. terrestris (Vauch.), Lyng.

M, under water, edge of Spot Pond.

FLORIDEÆ.

CHANTRANSIA, Fries.

C. Hermanni (Roth), Desv.

M, Spot Pond brook, Virginia Wood, in rapid current or at falls.

BATRACHOSPERMUM, Roth.†

B. Boryanum, Sirdt.

M, common in brooks in spring.

B. pyramidale, Sirdt.

M, brook by Highland Ave., Medford.

B. radians, Sirdt.

M, Spot Pond brook, Virginia Wood.

B. virgatum, Sirdt.

M, Spot Pond brook, Virginia Wood, chiefly, in spring, like the other species of this genus.

† B. Decaisneanum, Sirdt., has been found just outside the Fells Reservation limits in a spring in a hollow between Fulton and Elm Sts. It is the only recorded station in America.

LEMANEA, Bory.

L. fucina, Bory, var. **subtilis** (Sirdt.), Atk.

M, Cascade, growing on sloping or perpendicular rocks, in falling water or very swift current.

Division III. LICHENES.†
(By Clara E. Cummings.)

THELOSCHISTES, Norm.

T. lychneus, Nyl.
On trees; rare. $BB*$.

PARMELIA, Ach.

P. perlata (L.), Ach.
On trees; rather common. $M*$.

P. Borreri, Turn., var. **rudecta,** Tuck.
On trees; common. $M*$ and $BB*$.

P. saxatilis (L.), Fries, var. **sulcata,** Nyl.
On rocks. $M*$.

P. caperata (L.), Ach.
On trees; common. $M*$ and $BB*$.

P. conspersa (Ehrh.), Ach.
On rocks; common. $B*$, Great Blue Hill:—$M*$ and $BB*$.

PHYSCIA, DC.

P. stellaris, L.
On trees; common. $M*$ and $BB*$.

P. tribacia (Ach.), Tuck.
On trees; common. $BB*$.

P. hispida (Schreb., Fries), Tuck.
On trees; rare. $BB*$, with fibrils poorly developed.

† As but few localities are recorded, most of the letters indicating the Reservations where the species have been collected, follow the general habitat without comment.

P. obscura (Ehrh.), Nyl.
On trees; common. *M** and *BB**.

Var. endochrysea, Nyl.
On trees; rare. *BB**.

UMBILICARIA, Hoffm.

U. Dillenii, Tuck.
On rocks; common. *M** and *BB**.

U. pustulata (L.), Hoffm., var. **papulosa**, Tuck.
On rocks; common. *B**, *M** and *BB**.

STICTA, Schreb.

S. pulmonaria (L.), Ach.
On trees; common. *B**, Great Blue Hill: — *M**.

PELTIGERA, Willd.

P. horizontalis (L.), Hoffm.
On earth. *M**.

P. rufescens (Neck.), Hoffm.
On earth. *M**.

P. canina (L.), Hoffm.
On earth. *M**.

LECANORA, Ach.

L. varia (Ehrh.), Nyl.
On trees. *BB**.

RINODINA, Mass.

R. oreina (Ach.), Mass.
On rocks. *BB**.

PERTUSARIA, DC.

P. multipuncta (Turn.), Nyl.
On trees. *BB**.

STEREOCAULON, Schreb.

S. paschale (L.), Fries.
On rock. *M**, rather a small form.

CLADONIA, Hoffm.

C. pyxidata (L.), Fries.
On earth; common. *M**, Pine Hill: — *BB**.

C. fimbriata (L.), Fries, var. **tubæformis**, Fries.
On earth. *M**.

C. gracilis (L.), Nyl.
On earth. *BB**.

Var. **verticillata**, Fries.
On earth. *M**.

C. furcata (Huds.), Fries.
On earth. *BB**.

Var. **crispata**, Flœrke.
On earth. *M**.

Var. **racemosa**, Flœrke.
On earth; common. *M**.

C. rangiferina (L.), Hoffm.
On earth; rare. *BB**.

Var. **sylvatica**, L.
On earth; common. *M** and *BB**.

C. uncialis (L.), Fries.
On earth. *BB**.

C. cornucopioides (L.), Fries.
On earth; rare. *M**.

C. macilenta (Ehrh.), Hoffm.
On dead wood; rare. *M**.

C. cristatella, Tuck.
On earth and dead wood; common. *M** and *BB**.

BUELLIA, De Not.

B. petræa (Flot., Kœrb.), Tuck. var. ?
*M**, on rock.

ENDOCARPON, Hedw.

E. hepaticum, Ach., var. **complicatum**, Schær.
On rock. *M**.

Division IV. FUNGI.

(By A. B. Seymour and Flora W. Patterson.)

USTILAGINEÆ.

Urocystis Hypoxis, Thaxter.
Rare. *S*, on *Hypoxis erecta*, L.; the finding of this species is of considerable interest as it has been reported but once before, when it was collected by Dr. Roland Thaxter in Connecticut.

PHYCOMYCETES.

Synchytrium Myosotidis, var. **Potentillæ**, Farl.
Frequent. *M*, on *Potentilla Canadensis*, L.

Peronospora Corydalis, DBy.
Probably frequent. *B*, on *Corydalis glauca*, Pursh., near Great Blue Hill.

P. Ficariæ, Tul.
Common. *BB*, on *Ranunculus bulbosus*, L.

P. Linariæ, Fckl.
Frequent, but obscure. *S*, on *Linaria Canadensis*, Spreng.

PYRENOMYCETES.

Plowrightia morbosa (Schw.), Sacc.
Common. *BB*, on *Prunus Virginiana*, L.

Fungi Imperfecti.

Glœosporium Canadense, Ell. & Ev.
Common and very destructive. *B* and *M*, on *Quercus alba*, L.

Phyllosticta phomiformis, Sacc.
Frequent. *B*, on *Quercus alba*, L.

DISCOMYCETES.

Geoglossum Peckianum, Cooke.
Common. *M*, near Shiner Pool.

Leotia chlorocephala, Schw.
Common. *M*, near Shiner Pool.

Peziza hirta, Schum.
Common. *M*, on ground.

Rhytisma Solidaginis, Schw.
Frequent. *B*, on *Aster corymbosus*, Ait., Great Dome.

EXOASCEÆ.

Exoascus, sp.?
Very rare. *B*, on *Alnus incana*, Willd.; this is the first *Exoascus* found in America upon *Alnus* leaves, and the determination of the species is reserved for further collections and more critical study.

Taphrina Alni-incanæ (Kühn), Magnus.
Common. *M* and *B B*, on *Alnus incana*, Willd.

T. cærulescens (Mont. & Desm.), Tul.
Common. *M*, on *Quercus coccinea*, Wang., var. *tinctoria*, Gray.

T. flava, Farl.
Rare. *S*, on *Betula populifolia*, Ait.

Magnusiella Potentillæ (Farl.), Sadebeck.
Common. *B B*, on *Potentilla Canadensis*, L.

UREDINEÆ.

Cæoma nitens, Schw.
Abundant. *B*, on *Rubus villosus*, Ait.

Chrysomyxa pyrolata, Wint.
Scarce. *B*, on *Pyrola rotundifolia*, L., near Babel Rock.

Puccinia Anemones, P.
Frequent. *M*, on *Anemone nemorosa*, L.

P. graminis, Pers.
Everywhere. *B*, on *Agrostis alba*, L., var. *vulgaris*, Thurb.

Gymnosporangium clavipes, C. & P.
Frequent. *M*, on *Juniperus Virginiana*, L.

Rœstelia Botryapites, B.
Often abundant, but restricted. *B*, on *Amelanchier Canadensis*, Torr. & Gray, var. *oblongifolia*, Torr. & Gray, top of Hancock Hill.

R. globosa, Thaxter.
Frequent. *S*, on *Cratægus coccinea*, L.

BASIDIOMYCETES.

Cantharellus cibarius, Fr.
Common, edible. *M*, on ground.

Craterellus cornucopioides (L.), Pers.
On ground, frequent. *M*.

Exobasidium Vaccinii (Fckl.), Wor.
Common. *S*, on *Vaccinium vacillans*, Soland.

Hydnum suaveolens, Scop.
Common. *M*, on wood.

Polyporus versicolor (L.), Fr.
Common. *M*, on wood.

Schizophyllum commune, Fr.
Common. *M*, on wood.

INDEX.

GENERA AND COMMON NAMES.

Acalypha	68	Apple	27	Batrachospermum	132
Acer	17	Wild Balsam	32	Hayberry	71
Achillea	45	Aquilegia	5	Bearberry	52
Acorus	85	Arabis	8	Bedstraw	37
Actæa	6	Aralia	31	Beech	75
Adam's Needle	81	Arbutus, Trailing	52	Blue	78
Adder's Mouth	77	Arctium	47	Beechdrops	61
Adiantum	109	Arctostaphylos	52	Beggar-ticks, Common	45
Æsculus	17	Arenaria	12	Swamp	45
Agrimonia	26	Arethusa	78	Bellflower	50
Agrimony	26	Arisæma	85	Bellwort	82
Agropyrum	105	Aristida	99	Common	82
Agrostis	101	Arrowhead	86	Benjamin Bush	68
Ailanthus	15	Arrow Wood	35	Berberis	6
Alder	72	Artemisia	46	Betula	71
Black	15	Arum, Arrow	85	Bidens	45
White	53	Water	85	Bindweed	58
Alisma	86	Asclepias	56	Hiack	67
Allium	80	Ash	55	Birch	71
Allspice, Wild	68	European Mountain	27	Bitter-Nut	71
Alnus	72	Prickly	15	Bittersweet	59
Alopecurus	101	Asparagus	81	Climbing	16
Alsike Clover	19	Aspen	76	Shrubby	16
Alyssum	8	Aspidium	110	Blackberry	24
Amaranth	65	Asplenium	109	Black-eyed Susan	44
Amarantus	65	Asprella	105	Black Gum	35
Ambrosia	44	Aster	41	Bladderwort	61
Amelanchier	28	White-topped	41	Blood-root	7
Ampelopsis	17	Atrichum	121	Blue Beech	73
Amphicarpæa	22	Aulacomnium	121	Blueberry	51
Anabæna	128	Avena	102	Blue Curls	62
Anaphalis	44	Avens	25	Blue Flag	79
Andreæa	116	Azalea	53	Blue Joint	102
Andromeda	52			Bluets	36
Andropogon	99	Bachelor's Button	47	Blue Weed	58
Anemone	3	Balm of Gilead	77	Bœhmeria	69
Star	54	Balsam	15	Boneset, Upland	89
Anemonella	4	Balsam Apple, Wild	32	Botrychium	113
Anomodon	123	Baneberry	6	Bouncing Bet	11
Antennaria	43	Baptisia	19	Brachyelytrum	100
Anthemis	45	Barbarea	8	Brake	109
Anthoxanthum	99	Barberry	6	Brasenia	6
Anychia	65	Barbula	118	Brassica	8
Aphanochæte	131	Bartonia	57	Brier, Sweet	26
Aphyllon	61	Bartramia	120	Bromus	104
Apios	22	Basil	63	Broom-rape, Naked	61
Apocynum	56	Basswood	14	Brunella	64

INDEX.

Bryum	120	Checkerberry	52	Cucumber Root, Indian	82
Buckbean	57	Cheeses	13	Cudweed	44
Buckeye	17	Chelidonium	7	Currant	29
Buckthorn	16	Chelone	59	Cuscuta	58
Buckwheat	68	Chenopodium	65	Cylindrothecium	123
Climbing False	67	Cherry	28	Cynoglossum	57
Buda	12	Chess	104	Cyperus	88
Buellia	135	Chestnut	74	Cypress	68, 106
Bugleweed	63	Chickweed	12	Cypripedium	79
Bugloss, Viper's	58	Forked	65	Cystopteris	112
Bulbochæte	131	Indian	32		
Bulrush	89	Mouse-ear	12	Dactylis	103
Bunchberry	34	Chicory	48	Daisy	46
Burdock	47	Chimaphila	53	Ox-eye	46
Bur Marigold	45	Chokeberry	27	Dandelion	49
Bur-reed	85	Chrysanthemum	46	Dwarf	47
Butter-and-Eggs	59	Chrysomyxa	138	Fall	48
Buttercup	4	Chrysopogon	99	Dangleberry	51
Butternut	70	Chrysosplenium	28	Danthonia	102
Butterweed	43	Cicely, Sweet	33	Daucus	32
Button Bush	37	Cichorium	48	Day Lily	80
Buttonwood	70	Cicuta	33	Decodon	31
Buxbaumia	122	Cinna	101	Delphinium	5
		Cinquefoil	25	Deschampsia	102
Cæoma	137	Circæa	32	Desmodium	20
Calamagrostis	102	Cladonia	135	Deutzia	28
Calla	85	Clathrocystis	126	Dewberry	25
Callitriche	30	Clearweed	69	Dianthus	11
Calopogon	78	Cleavers	37	Dichelyma	122
Calothrix	127	Clematis	3	Dicksonia	112
Caltha	5	Clethra	53	Dicranella	117
Campanula	50	Climacium	123	Dicranum	117
Campion	11	Clover	19	Diervilla	36
Cancer-root	61	Bush	21	Diphyscium	122
Cantharellus	138	Sweet	20	Dock	66
Capsella	9	Club Moss	114	Dockmackie	35
Caraway	33	Cnicus	47	Dodder	58
Cardamine	7	Cockle	11	Dogbane	56
Cardinal Flower	50	Cocklebur	44	Dogwood	18, 34
Carex	90	Cœlosphærium	126	Doorweed	66
Carpet Weed	32	Coleochæte	132	Draparnaldia	131
Carpinus	73	Columbine	5	Dropwort	24
Carrion Flower	80	Comandra	68	Drosera	30
Carrot	32	Cone Flower	44	Drummondia	119
Carum	33	Conomitrium	117	Duckweed	86
Carya	70	Conopholis	61	Dulichium	88
Cassandra	52	Convallaria	81		
Castanea	74	Convolvulus	58	Eatonia	102
Catalpa	62	Coptis	5	Echinocystis	32
Cat-brier	80	Corallorhiza	77	Echium	58
Catchfly	11	Coral-root	77	Eglantine	26
Catnip	63	Corn Cockle	11	Elatine	13
Cat-tail, Common	84	Cornel	34	Elder	35
Ceanothus	16	Cornflower	47	Eleocharis	88
Cedar, Red	107	Cornus	34	Elm	69
White	106	Corydalis	7	Elodes	13
Celandine	7	Corylus	72	Elymus	105
Celastrus	16	Cow-lily	6	Endocarpon	136
Centaurea	47	Cow Wheat	61	Epigæa	52
Cephalanthus	37	Cranberry	51	Epilobium	31
Cerastium	12	Cranesbill	14	Epiphegus	61
Ceratodon	118	Cratægus	27	Equisetum	108
Chætophora	130	Craterellus	138	Eragrostis	103
Chamæcyparis	106	Cress, Bitter	7	Erechtites	46
Chamomile	45	Marsh	8	Erigeron	43
Chantransia	132	Rock	8	Eriocaulon	87
Chara	126	Spring	7	Eriophorum	89
Charlock	8	Water	8	Erythronium	82
Jointed	9	Winter	8	Eupatorium	38
Cheat	104	Crowfoot	4	Euphorbia	68

INDEX. 141

Everlasting	43, 44	Grass, Barnyard	97	Hellebore, False	83
Exoascus	137	Beard	99	Hemerocallis	80
Exobasidium	138	Bent	101	Hemlock	106
		Black Oat	99	Water	33
Fagopyrum	68	Blue-eyed	79	Hemp Weed, Climbing	38
Fagus	75	Bottle Brush	105	Hepatica	3
Featherfoil	54	Bristly Foxtail	98	Herb Robert	14
Fern, Beech	110	Brome	104	Hickory	70
Bladder	112	Brown Bent	101	Hieracium	48
Chain	109	Cat's Tail	100	Hierochloe	99
Cinnamon	113	Cotton	89	Hoary Pea	20
Flowering	112	Crab	96	Hog Peanut	22
Grape	113	Drop-seed	100, 101	Holcus	102
Lady	110	Feather	99	Holly	15
Rattlesnake	113	Fescue	104	Mountain	16
Sensitive	112	Fowl Meadow	103	Honeysuckle	36
Shield	110	Foxtail	98, 101	Bush	36
Sweet	71	Green Foxtail	98	White Swamp	53
Wood	111	Hair	101	Horehound, Water	63
Festuca	104	Herd's	100	Hornbeam	73
Fever Bush	68	June	103	Hop	73
Feverfew	46	Kentucky Blue	103	Horse-brier	80
Feverwort	36	Lyme	105	Horse-Chestnut	17
Fimbristylis	89	Manna	103	Horse Gentian	36
Fireweed	31, 46	Meadow	103	Horsetail	108
Fissidens	117	Meadow Soft	102	Horseweed	43
Five-finger	25	Orchard	103	Hottonia	54
Flag, Blue	79	Panic	96	Hound's Tongue	57
Cat-tail	84	Poverty	99	Houstonia	36
Sweet	85	Rattlesnake	103	Huckleberry	51
Flax	14	Reed Bent	102	Hydnum	138
Toad	59	Rice Cut	98	Hydrangea	28
Fleabane	43	Spear	103	Hydrocotyle	33
Daisy	43	Star	80	Hypericum	12
Floating Heart	57	Sweet	99	Hypnum	123
Flower-de-Luce	79	Sweet Vernal	99	Hypoxis	80
Fontinalis	122	Thin	101	Hyssop, Hedge	60
Forget-me-not	57	Triple-awned	99		
Foxglove, False	60	Velvet	102	Ilex	15
Fragaria	25	White	98	Ilysanthes	60
Fraxinus	55	Wild Oat	102	Impatiens	15
Frost-weed	9	Wire	103	Indian Bean	62
Funaria	120	Witch	105	Indian Cucumber Root	82
Funkia	81	Wood Reed	101	Indian Hemp	56
		Yellow-eyed	83	Indian Pipe	54
Galeopsis	64	Zebra	98	Indian Tobacco	50
Gale, Sweet	71	Gratiola	60	Indian Turnip	85
Galingale	88	Graveyard Flower	68	Indigo, False	19
Galium	37	Greenbrier	80	Wild	19
Garget	66	Grimmia	118	Inkberry	16
Garlic	80	Ground Nut	22	Innocence	36
Gaultheria	52	Ground Pine	114	Ipomœa	58
Gaylussacia	51	Groundsel	46	Iris	79
Gentian, Horse	36	Gymnosporangium	138	Iron Wood	73
Geoglossum	137			Isoetes	114
Geranium	14	Habenaria	78	Ivy, Ground	63
Gerardia	60	Hamamelis	30	Poison	18
Geum	25	Hapalosiphon	128		
Gill-over-the-ground	63	Hardhack	23		
Glœocapsa	126	Hassallia	128	Jack-in-the-Pulpit	85
Glœosporium	136	Hawkbit	48	Jewel Weed	15
Glyceria	103	Hawkweed	48	Joe-Pye Weed	38
Gnaphalium	44	Hawthorn	27	Juglans	70
Goat's Rue	20	Hazel-Nut	72	Juncus	83
Golden Rod	39	Hazel, Witch	30	June Berry	28
Goldthread	5	Hedeoma	63	Juniper	107
Goodyera	78	Hedge Hyssop	60	Juniperus	107
Gooseberry	29	Hedwigia	119		
Goosefoot, Maple-leaved	65	Helianthemum	9	Kalmia	53
Grape	17	Helianthus	44	Kinnikinnik	35

Knapweed	47	Maidenhair	109	New Jersey Tea	16
Knawel	65	False	109	Nightshade	59
Knotweed	66	Mallow	13	Enchanter's	32
Krigia	47	Malva	13	Nitella	125
		Maple	17	Nostoc	128
Lactuca	49	Marigold Bur	45	Nuphar	6
Ladies' Tresses	78	Marsh	5	Nymphæa	6
Lady's Slipper	79	Water	45	Nyssa	35
Lady's Thumb	67	Marsh Marigold	5		
Lambkill	53	Mayflower	52	Oak	73
Larkspur	5	Mayweed	45	Poison	18
Laurel, American	53	Meadow Beauty	31	Oakesia	82
Mountain	53	Meadow Rue	4	Oat	102
Sheep	53	Meadow Sweet	23	Œdogonium	131
Leather Leaf	52	Medeola	82	Œnothera	32
Lecanora	134	Medicago	20	Onion	80
Lechea	10	Medick	20	Onoclea	112
Leersia	98	Melampyrum	61	Ophiocytium	130
Lemanea	133	Melilot	20	Orange-grass	13
Lemna	86	Melilotus	20	Orchis, Ragged-fringed	79
Leontodon	48	Mentha	62	White-fringed	79
Leonurus	64	Menyanthes	57	Ornithogalum	80
Leotia	137	Mercury, Three-seeded	68	Orpine	29
Lepidium	9	Mermaid Weed	30	Orthotrichum	119
Leptotrichum	118	Mexican Tea	65	Oryzopsis	100
Leskea	123	Microstylis	77	Oscillatoria	127
Lespedeza	21	Mignonette	9	Osmorrhiza	33
Lettuce	49	Mikania	38	Osmunda	112
White	48	Milfoil, Water	30	Ostrya	73
Leucobryum	118	Milkweed	56	Oxalis	15
Leucodon	122	Milkwort	18	Ox-eye Daisy	46
Leucothoe	52	Millet, True	97		
Lever Wood	73	Mimulus	59	Pæonia	6
Life-of-Man	34	Mint	62	Panicum	96
Ligustrum	56	Mountain	63	Parmelia	133
Lilac	56	Miscanthus	98	Parsnip	32
Lilium	82	Mitchella	37	Water	33
Lily	82	Mnium	121	Partridge Berry	37
Day	80	Mocker-Nut	70	Paspalum	96
Lily of the Valley	81	Mock Orange	29	Pastinaca	32
Limnanthemum	57	Mollugo	32	Peach	23
Linaria	59	Monoses	54	Pea, Hoary	20
Linden	14	Moneywort	55	Peanut, Hog	22
Lindera	68	Monkey Flower	59	Pear	27
Linum	14	Monotropa	54	Peat Moss	115
Liparis	77	Moonwort	113	Pediastrum	130
Liquorice, Wild	37	Morning Glory	58	Pedicularis	61
Live-for-ever	29	Wild	58	Peltandra	85
Liver-leaf	3	Morus	69	Peltigera	134
Lobelia	50	Motherwort	64	Pennyroyal, Bastard	62
Locust, Common	20	Mougeotia	129	Mock	63
Locust-tree	20	Mugwort	46	Pennywort, Water	33
Lonicera	36	Muhlenbergia	100	Penthorum	29
Loosestrife	31, 55	Mulberry	69	Peony	6
False	31	Mullein	59	Pepperbush, Sweet	53
Swamp	31	Mustard, Black	9	Peppergrass	9
Lousewort	61	Hedge	8	Pepperidge	35
Love Apple	58	Myosotis	57	Pepper, Water	67
Ludwigia	31	Myrica	71	Peronospora	136
Luzula	84	Myriophyllum	30	Pertusaria	134
Lychnis	11			Petunia	59
Lycopersicum	58	Naiad	87	Peziza	137
Lycopodium	114	Naias	87	Phegopteris	110
Lycopus	63	Nasturtium	8	Philadelphus	29
Lyngbya	127	Nemopanthes	16	Philonotis	120
Lysimachia	55	Nepeta	63	Phleum	100
Lythrum	31	Nettle	69	Phormidium	127
		False	69	Phyllosticta	137
Magnusiella	137	Hedge	64	Physcia	133
Maianthemum	81	Hemp	64	Physcomitrium	120

INDEX. 143

Phytolacca	66	Quercus	73	Sedge	90
Picea	106	Quillwort	114	Sedum	29
Pickerel-weed	83	Quince	27	Selaginella	114
Pigeon Berry	66			Self-heal	64
Pig-Nut	70	Racomitrium	119	Senecio	46
Pigweed	65	Radish	9	Sericocarpus	41
Amaranth	65	Ragweed	44	Service Berry	28
Pilea	69	Ragwort, Golden	46	Setaria	98
Pimpernel, False	60	Ranunculus	4	Shadbush	28
Pine	105	Raphanus	9	Sheep's Fescue	104
Pinesap	54	Raspberry	24	Shepherd's Purse	9
Pine-weed	13	Rattlesnake-root	48	Shinleaf	54
Pink	11	Rattlesnake-weed	48	Sickle-pod	8
Wild	11	Red Top	101	Side-saddle Flower	7
Pinus	105	False	103	Silene	11
Pinweed	10	Reseda	9	Sisymbrium	8
Pipewort	87	Rhamnus	16	Sisyrinchium	79
Pipsissewa	53	Rhexia	31	Sium	33
Pitcher-plant	7	Rhododendron	53	Skullcap	64
Plantago	64	Rhodora	53	Skunk Cabbage	85
Plantain	64	Rhus	18	Smartweed, Common	67
Rattlesnake	78	Rhynchospora	90	Smilacina	81
Robin's	43	Rhytisma	137	Smilax	80
Water	86	Ribes	29	Snake-head	59
Plantanus	70	Ribgrass	64	Snakeroot, Black	34
Plectonema	127	Rice, Mountain	100	White	39
Plowrightia	136	Richweed	69	Sneezewort	45
Plum	23	Rinodina	134	Soapwort	11
Poa	103	Robinia	20	Solanum	59
Pogonatum	121	Rocket, Yellow	8	Solidago	39
Pogonia	78	Rock-Rose	9	Solomon's Seal	81
Poison Ivy	18	Rœstelia	138	Dwarf	81
Poke, Common	66	Roman Wormwood	44	False	81
Indian	83	Rosa	26	Sonchus	49
Pokeweed	66	Rose	26	Sorrel	66
Polygala	18	Rose Bay	53	Field	66
Polygonatum	81	Rubus	24	Wood	15
Polygonum	66	Rudbeckia	44	Sour-gum Tree	35
Polypodium	109	Rue-Anemone	4	Spanish Bayonet	81
Polypody	109	Rue, Goat's	20	Sparganium	85
Polyporus	138	Meadow	4	Spatter Dock	6
Polytrichum	121	Rumex	66	Specularia	50
Pond-lily, Yellow	6	Rush	83	Speedwell	60
Pondweed	87	Beak	90	Sphagnum	115
Pontederia	83	Club	89	Spice Bush	68
Poplar	76	Scouring	108	Spikenard	34
Populus	76	Spike	88	False	81
Portulaca	12	Wood	84	Spiræa	23
Potamogeton	87	Rye, Wild	105	Spiranthes	78
Potentilla	25			Spirodela	86
Pottia	118	Sagittaria	86	Spirogyra	129
Prenanthes	48	Salix	75	Spleenwort	109
Prickly Ash	15	Sambucus	35	Sporobolus	101
Primrose, Evening	32	Sand-Spurrey	12	Spruce	106
Prince's Pine	53	Sandwort	12	Spurge	68
Privet	56	Sanguinaria	7	Squaw-root	61
Proserpinaca	30	Sanicula	34	Stachys	64
Prunus	23	Saponaria	11	Staff-tree	16
Pteris	109	Sarracenia	7	Star-flower	54
Puccinia	138	Sarsaparilla, Wild	34	Star-of-Bethlehem	80
Purslane	12	Sassafras	68	Starwort	12
Water	31	Savin	107	Water	30
Pusley	12	Saxifraga	28	Steeple Bush	23
Pycnanthemum	63	Saxifrage	28	Steironema	54
Pylaisia	123	Golden	28	Stellaria	12
Pyrola	54	Schizophyllum	138	Stereocaulon	134
One-flowered	54	Schizothrix	126	Sticta	134
Pyrus	27	Scirpus	89	Stigeoclonium	131
		Scleranthus	65	Stigonema	128
Quaker Ladies	36	Scutellaria	64	Stipa	99

St. John's-wort	13	Toad Flax	59	Virgin's Bower	3
Marsh	13	Bastard	68	Vitis	17
Stonecrop	29	Tolypothrix	128	Volvox	130
Ditch	29	Tomato	58		
Strawberry	25	Touch-me-not, Spotted	15	Wake Robin	82
Succory	48	Tree of Heaven	15	Walnut	70
Sugar Plum	28	Trefoil	19	Water-Crowfoot	4
Sumach	18	Tick	20	Water Hemlock	38
Chinese	15	Trentepohlia	131	Water-lily	6
Sundew	30	Trichostema	62	Water Marigold	45
Sunflower	44	Trientalis	54	Water Milfoil	30
Swamp Honeysuckle,		Trifolium	19	Water Shield	6
White	53	Trillium	82	Water Violet	54
Sweet Brier	26	Triosteum	36	Waterwort	13
Sweet Cicely	33	Triticum	105	Wax-work	16
Sweet Fern	71	Tsuga	106	Webera	120
Sweet Flag	85	Tupelo	35	Weisia	116
Sweet Gale	71	Turnip	9	Wheat	105
Sycamore	70	Turtle-head	59	Cow	61
Symploca	127	Twayblade	77	False	105
Symplocarpus	85	Typha	84	White-weed	46
Synchytrium	136			Wild Bean	22
Syringa	29, 56	Ulmus	69	Wild Oats	82
		Ulota	119	Willow	75
		Ulothrix	130	Willow Herb	31
Tanacetum	46	Umbilicaria	134	Wind-flower	3
Tansy	46	Urocystis	136	Winterberry, Smooth	16
Taphrina	137	Urtica	69	Wintergreen	53
Taraxacum	49	Utricularia	61	Aromatic	52
Tare	22	Uvularia	82	Creeping	52
Tear-thumb	67			Witch Hazel	30
Tephrosia	20			Withe-rod	36
Tetraphis	119	Vaccinium	51	Wood Betony	61
Tetraspora	130	Vaucheria	132	Woodbine	17
Thalictrum	4	Venus's Looking-glass	50	Woodsia	112
Thelia	122	Veratrum	83	Woodwardia	109
Theloschistes	133	Verbascum	59	Wormwood	46
Thimbleberry	24	Verbena	62	Roman	44
Thistle	47	Veronica	60		
Sow	49	Vervain	62	Xanthium	44
Star	47	Vetch	22	Xanthoxylum	15
Thorn	27	Viburnum	35	Xyris	83
Thoroughwort	38	Vicia	22		
Three-seeded Mercury	68	Viola	10	Yarrow	45
Thyme	63	Violet	10	Yellow Rocket	8
Thymus	63	Dog-tooth	82	Yucca	81
Tick Trefoil	20	Water	54		
Tilia	14	Viper's Bugloss	58	Zygnema	128
Timothy	100	Virginian Creeper	17	Zygogonium	129

CORRECTIONS.

Page 15, line 9. After BALSAM insert a period.
" 27, " 28. For Hawthorn read HAWTHORN.
" 54, " 16. After Hypopitys insert L.
" 61, " 19 and 20. For WALLROTH read Wallroth.

www.ingramcontent.com/pod-product-compliance
Lightning Source LLC
Chambersburg PA
CBHW022124160426
43197CB00009B/1146